363.73   Fisher, Marshall.
Fis
          The ozone layer.

                                        20476
$19.95

DATE			

THE OZONE
LAYER

EARTH • AT • RISK

EARTH • AT • RISK

THE OZONE LAYER

by Marshall Fisher

Introduction by
Russell E. Train

Chairman of the Board of
Directors,
World Wildlife Fund and
The Conservation Foundation

CHELSEA HOUSE PUBLISHERS

new york   philadelphia

CHELSEA HOUSE PUBLISHERS
EDITOR-IN-CHIEF: Remmel Nunn
MANAGING EDITOR: Karyn Gullen Browne
COPY CHIEF: Mark Rifkin
PICTURE EDITOR: Adrian G. Allen
ART DIRECTOR: Maria Epes
ASSISTANT ART DIRECTOR: Noreen Romano
MANUFACTURING MANAGER: Gerald Levine
SYSTEMS MANAGER: Lindsey Ottman
PRODUCTION MANAGER: Joseph Romano
PRODUCTION COORDINATOR: Marie Claire Cebrián

EARTH AT RISK
SENIOR EDITOR: Jake Goldberg

Staff for *The Ozone Layer*
ASSOCIATE EDITOR: Karen Hammonds
SENIOR COPY EDITOR: Laurie Kahn
EDITORIAL ASSISTANT: Ian Wilker
PICTURE RESEARCHER: Diana Gongora
DESIGNER: Maria Epes
LAYOUT: Marjorie Zaum

3 5 7 9 8 6 4

Library of Congress Cataloging-in-Publication Data
Fisher, Marshall John.
The ozone layer/Marshall John Fisher; introduction by Russell E.
Train.
   p. cm.—(Earth at risk)
Includes bibliographical references and index.
Summary: The ozone layer protects life on earth from harmful
ultraviolet radiation, but this protective shield is being damaged
by chlorofluorocarbons and other pollutants that are now being
generated on the earth.
ISBN 0-7910-1576-9
       0-7910-1601-3 (pbk.)
1. Ozone layer—Juvenile literature. 2. Chlorofluorocarbons—
Environmental aspects—Juvenile literature. 3. Environmental
protection—Juvenile literature. 4. Man—Influence on nature—
Juvenile literature. [Ozone layer. 2. Chlorofluorocarbons.]
I. Title. II. Series.   91-15913
QC881.2.09F57 1992   CIP
363.73'926—dc20   AC

C O N T E N T S

INTRODUCTION

Russell E. Train

Administrator, Environmental Protection Agency, 1973 to
1977; Chairman of the Board of Directors, World Wildlife
Fund and The Conservation Foundation

There is a growing realization that human activities increasingly
are threatening the health of the natural systems that make life possible
on this planet. Humankind has the power to alter nature fundamentally,
perhaps irreversibly.

This stark reality was dramatized in January 1989 when *Time*
magazine named Earth the "Planet of the Year." In the same year, the
Exxon *Valdez* disaster sparked public concern over the effects of human
activity on vulnerable ecosystems when a thick blanket of crude oil
coated the shores and wildlife of Prince William Sound in Alaska. And,
no doubt, the 20th anniversary celebration of Earth Day in April 1990
renewed broad public interest in environmental issues still further. It is
no accident then that many people are calling the years between 1990
and 2000 the "Decade of the Environment."

And this is not merely a case of media hype, for the 1990s will
truly be a time when the people of the planet Earth learn the meaning of
the phrase "everything is connected to everything else" in the natural
and man-made systems that sustain our lives. This will be a period when
more people will understand that burning a tree in Amazonia adversely
affects the global atmosphere just as much as the exhaust from the cars
that fill our streets and expressways.

Central to our understanding of environmental issues is the
need to recognize the complexity of the problems we face and the

relationships between environmental and other needs in our society. Global warming provides an instructive example. Controlling emissions of carbon dioxide, the principal greenhouse gas, will involve efforts to reduce the use of fossil fuels to generate electricity. Such a reduction will include energy conservation and the promotion of alternative energy sources, such as nuclear and solar power.

The automobile contributes significantly to the problem. We have the choice of switching to more energy efficient autos and, in the longer run, of choosing alternative automotive power systems and relying more on mass transit. This will require different patterns of land use and development, patterns that are less transportation and energy intensive.

In agriculture, rice paddies and cattle are major sources of greenhouse gases. Recent experiments suggest that universally used nitrogen fertilizers may inhibit the ability of natural soil organisms to take up methane, thus contributing tremendously to the atmospheric loading of that gas—one of the major culprits in the global warming scenario.

As one explores the various parameters of today's pressing environmental challenges, it is possible to identify some areas where we have made some progress. We have taken important steps to control gross pollution over the past two decades. What I find particularly encouraging is the growing environmental consciousness and activism by today's youth. In many communities across the country, young people are working together to take their environmental awareness out of the classroom and apply it to everyday problems. Successful recycling and tree-planting projects have been launched as a result of these budding environmentalists who have committed themselves to a cleaner environment. Citizen action, activated by youthful enthusiasm, was largely responsible for the fast-food industry's switch from rainforest to domestic beef, for pledges from important companies in the tuna industry to use fishing techniques that would not harm dolphins, and for the recent announcement by the McDonald's Corporation to phase out polystyrene "clam shell" hamburger containers.

Despite these successes, much remains to be done if we are to make ours a truly healthy environment. Even a short list of persistent issues includes problems such as acid rain, ground-level ozone and smog, and airborne toxins; groundwater protection and nonpoint sources of pollution, such as runoff from farms and city streets; wetlands protection; hazardous waste dumps; and solid waste disposal, waste minimization, and recycling.

Similarly, there is an unfinished agenda in the natural resources area: effective implementation of newly adopted management plans for national forests; strengthening the wildlife refuge system; national park management, including addressing the growing pressure of development on lands surrounding the parks; implementation of the Endangered Species Act; wildlife trade problems, such as that involving elephant ivory; and ensuring adequate sustained funding for these efforts at all levels of government. All of these issues are before us today; most will continue in one form or another through the year 2000.

Each of these challenges to environmental quality and our health requires a response that recognizes the complex nature of the problem. Narrowly conceived solutions will not achieve lasting results. Often it seems that when we grab hold of one part of the environmental balloon, an unsightly and threatening bulge appears somewhere else.

The higher environmental issues arise on the national agenda, the more important it is that we are armed with the best possible knowledge of the economic costs of undertaking particular environmental programs and the costs associated with not undertaking them. Our society is not blessed with unlimited resources, and tough choices are going to have to be made. These should be informed choices.

All too often, environmental objectives are seen as at cross-purposes with other considerations vital to our society. Thus, environmental protection is often viewed as being in conflict with economic growth, with energy needs, with agricultural productions, and so on. The time has come when environmental considerations must be fully integrated into every nation's priorities.

One area that merits full legislative attention is energy efficiency. The United States is one of the least energy efficient of all the industrialized nations. Japan, for example, uses far less energy per unit of gross national product than the United States does. Of course, a country as large as the United States requires large amounts of energy for transportation. However, there is still a substantial amount of excess energy used, and this excess constitutes waste. More fuel efficient autos and home heating systems would save millions of barrels of oil, or their equivalent, each year. And air pollutants, including greenhouse gases, could be significantly reduced by increased efficiency in industry.

I suspect that the environmental problem that comes closest to home for most of us is the problem of what to do with trash. All over the world, communities are wrestling with the problem of waste disposal. Landfill sites are rapidly filling to capacity. No one wants a trash and garbage dump near home. As William Ruckelshaus, former EPA administrator and now in the waste management business, puts it, "Everyone wants you to pick up the garbage and no one wants you to put it down!"

At the present time, solid waste programs emphasize the regulation of disposal, setting standards for landfills, and so forth. In the decade ahead, we must shift our emphasis from regulating waste disposal to an overall reduction in its volume. We must look at the entire waste stream, including product design and packaging. We must avoid creating waste in the first place. To the greatest extent possible, we should then recycle any waste that is produced. I believe that, while most of us enjoy our comfortable way of life and have no desire to change things, we also know in our hearts that our "disposable society" has allowed us to become pretty soft.

Land use is another domestic issue that might well attract legislative attention by the year 2000. All across the United States, communities are grappling with the problem of growth. All too often, growth imposes high costs on the environment—the pollution of aquifers; the destruction of wetlands; the crowding of shorelines; the loss of wildlife habitat; and the loss of those special places, such as a historic structure or area, that give a community a sense of identity. It is worth noting that growth is not only the product of economic development but of population movement. By the year 2010, for example, experts predict that 75% of all Americans will live within 50 miles of a coast.

It is important to keep in mind that we are all made vulnerable by environmental problems that cross international borders. Of course, the most critical global conservation problems are the destruction of tropical forests and the consequent loss of their biological capital. Some scientists have calculated extinction rates as high as 11 species per hour. All agree that the loss of species has never been greater than at the present time; not even the disappearance of the dinosaurs can compare to today's rate of extinction.

In addition to species extinctions, the loss of tropical forests may represent as much as 20% of the total carbon dioxide loadings to the atmosphere. Clearly, any international approach to the problem of global warming must include major efforts to stop the destruction of forests and to manage those that remain on a renewable basis. Debt for nature swaps, which the World Wildlife Fund has pioneered in Costa Rica, Ecuador, Madagascar, and the Philippines, provide a useful mechanism for promoting such conservation objectives.

Global environmental issues inevitably will become the principal focus in international relations. But the single overriding issue facing the world community today is how to achieve a sustainable balance between growing human populations and the earth's natural systems. If you travel as frequently as I do in the developing countries of Latin America, Africa, and Asia, it is hard to escape the reality that expanding human populations are seriously weakening the earth's resource base. Rampant deforestation, eroding soils, spreading deserts, loss of biological diversity, the destruction of fisheries, and polluted and degraded urban environments threaten to spread environmental impoverishment, particularly in the tropics, where human population growth is greatest.

It is important to recognize that environmental degradation and human poverty are closely linked. Impoverished people desperate for land on which to grow crops or graze cattle are destroying forests and overgrazing even more marginal land. These people become trapped in a vicious downward spiral. They have little choice but to continue to overexploit the weakened resources available to them. Continued abuse of these lands only diminishes their productivity. Throughout the developing world, alarming amounts of land rendered useless by over-grazing and poor agricultural practices have become virtual wastelands, yet human numbers continue to multiply in these areas.

From Bangladesh to Haiti, we are confronted with an increasing number of ecological basket cases. In the Philippines, a traditional focus of U.S. interest, environmental devastation is widespread as deforestation, soil erosion, and the destruction of coral reefs and fisheries combine with the highest population growth rate in Southeast Asia.

Controlling human population growth is the key factor in the environmental equation. World population is expected to at least double to about 11 billion before leveling off. Most of this growth will occur in the poorest nations of the developing world. I would hope that the United States will once again become a strong advocate of international efforts to promote family planning. Bringing human populations into a sustainable balance with their natural resource base must be a vital objective of U.S. foreign policy.

Foreign economic assistance, the program of the Agency for International Development (AID), can become a potentially powerful tool for arresting environmental deterioration in developing countries. People who profess to care about global environmental problems—the loss of biological diversity, the destruction of tropical forests, the greenhouse effect, the impoverishment of the marine environment, and so on—should be strong supporters of foreign aid planning and the principles of sustainable development urged by the World Commission on Environment and Development, the "Brundtland Commission."

If sustainability is to be the underlying element of overseas assistance programs, so too must it be a guiding principle in people's practices at home. Too often we think of sustainable development only in terms of the resources of other countries. We have much that we can and should be doing to promote long-term sustainability in our own resource management. The conflict over our own rain forests, the old growth forests of the Pacific Northwest, illustrates this point.

The decade ahead will be a time of great activity on the environmental front, both globally and domestically. I sincerely believe we will be tested as we have been only in times of war and during the Great Depression. We must set goals for the year 2000 that will challenge both the American people and the world community.

Despite the complexities ahead, I remain an optimist. I am confident that if we collectively commit ourselves to a clean, healthy environment we can surpass the achievements of the 1980s and meet the serious challenges that face us in the coming decades. I hope that today's students will recognize their significant role in and responsibility for bringing about change and will rise to the occasion to improve the quality of our global environment.

A satellite photograph of the earth, taken from an altitude of 22,300 miles. The planet and its atmosphere are more vulnerable to human activity than was once believed.

THE WORLD'S SECURITY BLANKET

Many miles above the ground, hovering over every nation and every person on earth, is a cloud of invisible gas called *ozone*. This gas serves as a sort of global security blanket because it absorbs ultraviolet radiation emitted by the sun and prevents it from reaching the earth's surface at full strength, where it would burn everything to a crisp and render life impossible. Even a small depletion of the ozone layer translates into dangerously high levels of ultraviolet radiation at ground level, which could damage plant and animal life and cause severe human health problems, including many thousands of new skin cancer cases each year. Human industrial activity—the production of ozone-destroying chemical compounds—is now causing such a depletion, which could eventually threaten life on earth.

The ozone layer is one of many atmospheric phenomena that together have created an environment on this planet suitable for life. The earth's atmosphere blocks out enough ultraviolet light to protect life while letting in enough radiation of other wavelengths to keep the earth warm and provide light with which to see. The atmosphere also serves as a thermal covering, with carbon dioxide (CO_2) and other gases trapping sufficient heat to keep the planet from freezing, yet letting enough escape so that it does not overheat.

To see what happens when these atmospheric conditions are slightly different, one need only look at the closest planets. Venus and Mars each have an atmosphere rich in carbon dioxide, but at the wrong concentration to sustain life as we know it. Venus has too much carbon dioxide, which traps too much heat and keeps the surface temperature well over the boiling point of water. Mars, on the other hand, has a thin atmosphere that allows too much heat to escape, resulting in a freezing-cold surface. Scientific surveys of these planets have shown fairly conclusively that both are devoid of life.

Earth is the only place, at least in this solar system, where life was able to evolve. The planet's size maintains a gravitational field neither too strong nor too weak, so that just the right sort of atmosphere is retained; this atmosphere, along with the earth's distance from the sun, helps to create the right temperatures. As a result, for the past 100 million years the earth has been a perfect setting in which life can prosper. But it is a mistake to think that the earth has always been as conducive to life as it is now or that it will necessarily remain so. The earth has existed for about 4.5 billion years, and for nearly 4 billion years, or almost 90% of that lifetime, it was devoid of complex life-forms. To understand just how precarious is the present atmospheric balance, it is helpful to

A high-resolution photograph of the surface of Mars, taken by the Viking Lander 2 in May 1979, shows rocks and soil coated with ice. A thin atmosphere results in frigid temperatures on the planet.

look at how that balance was created—in particular, how the ozone layer came to exist.

LOOKING BACK

Ozone is made from oxygen, but for billions of years there was virtually no oxygen in the earth's atmosphere. The earth's primitive atmosphere was probably composed of water vapor (which formed the oceans), carbon dioxide, sulfur dioxide, nitrogen, and traces of other gases released by volcanic and other geothermal activity from the planet's hot interior. Because there was no oxygen, there was no ozone, and all the sun's ultraviolet radiation reached the earth's surface. This ultraviolet radiation had enough energy to break up any complex molecules that may have formed and so prevented complex life from developing. At a certain depth underwater, however, just the right amount of ultraviolet radiation was screened out. It was there—in what is often called the "primordial soup"—that the same radiation that precluded life on land probably led to the creation of primitive life-forms.

These ancient oceans were packed with carbon and other organic materials that needed only a burst of energy in order to form living organisms. The ultraviolet light provided the spark, and the organic compounds fused first into complex, self-replicating molecules and ultimately into primitive single-celled organisms. These early life-forms fed off a process called *fermentation*. With the aid of energy from sunlight, the organisms broke down carbon dioxide dissolved in the water, using the carbon to grow. Later, around 2.7 billion years ago, organisms evolved the process known as *photosynthesis*, in which energy is derived from visible sunlight, with oxygen being the by-product. Eventually the atmosphere contained a considerable amount of oxygen. This element was poisonous to the first life-forms, but some organisms evolved a mechanism called *respiration*, in which energy is obtained from the reaction between oxygen and carbon, giving off carbon dioxide as a waste product. Respiration was a greater and quicker source of energy than photosynthesis and paved the way for the evolution of animal life. Eventually a healthy balance was achieved between oxygen-producing plants and carbon dioxide–producing animals. The atmosphere now contained a significant percentage of oxygen, and its appearance led to the development of the ozone layer.

FROM OXYGEN TO OZONE

A molecule of oxygen contains two oxygen atoms and is written O_2. A molecule of ozone contains three oxygen atoms—O_3. This is a subtle distinction, but one with critical consequences. Whereas oxygen is necessary for life as we know it, ozone in the *biosphere*—that part of the earth and its atmosphere inhabited by living things—is a poison. Animals cannot breathe it, but they do not normally have to, except when ozone "smog" is created at ground level by the action of sunlight upon air pollution.

The *natural* atmospheric ozone layer, which serves to protect life by absorbing radiation, is most concentrated between

Ozone molecules are formed from oxygen in the earth's stratosphere.

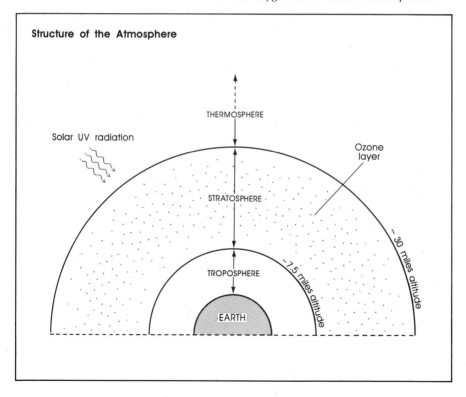

Structure of the Atmosphere

THERMOSPHERE

Solar UV radiation

Ozone layer

STRATOSPHERE

~ 30 miles altitude

TROPOSPHERE

~ 7.5 miles altitude

EARTH

12 and 20 miles (19 and 32 kilometers) above the earth, in a region of the earth's atmosphere called the *stratosphere*. Below the stratosphere, from the surface of the earth up to an average altitude of approximately 7.5 miles (12 km), is the *troposphere*—that part of the atmosphere that organisms breathe and in which most weather takes place. Above all that, above the portion that people recognize as "air," is where oxygen is turned into ozone in a process that absorbs the sun's ultraviolet radiation.

The sun radiates a wide spectrum of what is called *electromagnetic radiation*—wavelike oscillations, or vibrations, of electromagnetic fields that transmit energy through space. Most of the sun's radiation is visible light, with a wavelength of between 400 and 750 nanometers (nm; 1 nm = 1 billionth of a meter). But the sun also emits radiation that is invisible, with wavelengths either shorter or longer than what human eyes can detect. Examples of radiation with longer wavelengths include radio waves and infrared radiation. These are harmless because their longer wavelength carries less energy—the shorter the wavelength, the higher its energy, or radiation, content and the more dangerous it is to life on earth. Ultraviolet light has a shorter wavelength than visible light. Some ultraviolet radiation, measuring between 320 nm and 400 nm (called UV-A), is generally not considered harmful. But rays with wavelengths shorter than this are very destructive to life. UV-B, between 280 nm and 320 nm, is what leaks through the ozone layer, potentially causing skin cancer and other ailments. UV-C, with wavelengths measuring from 200 nm to 280 nm, would be an instant death ray were any of it to get through the ozone. There was a time when it did; not until the production of oxygen led to

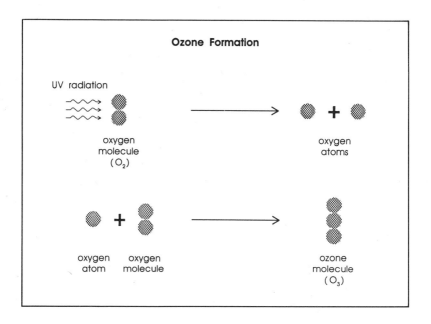

Ozone Formation

UV radiation

oxygen
molecule
(O_2)

oxygen
atoms

oxygen   oxygen
atom   molecule

ozone
molecule
(O_3)

the creation of the ozone layer, in fact, was life able to crawl out of the protective seas.

Ironically, ozone was originally and is still produced by the very ultraviolet radiation that threatens life. When a molecule of oxygen that floats up to the stratosphere is bombarded by ultraviolet rays, it splits into two free oxygen atoms, a reaction known as *photodissociation.* Each of these free atoms can then react with another oxygen molecule to form a molecule of ozone—three bonded oxygen atoms.

The reason ozone normally forms at a certain altitude is that above a certain height there are too few molecules to cause a significant number of the molecular collisions that produce ozone, whereas below a certain altitude less ultraviolet radiation penetrates to split oxygen molecules.

Under natural conditions, ozone exists in stable concentrations because at the same time that reactions turn oxygen into ozone, other reactions turn ozone back into oxygen. The oxygen-to-ozone reactions are sparked by UV-C radiation, which is powerful enough to break apart the strong bonds in O_2. UV-B light is less powerful but more abundant. It is this type that breaks up ozone and turns it back into oxygen, because the chemical bonds in ozone are weaker. In this case, a molecule of ozone is broken apart by ultraviolet energy into an oxygen molecule and a free oxygen atom. When ultraviolet light breaks up either oxygen or ozone molecules, its energy is absorbed and given off as heat, warming the stratosphere. In this way, all of the UV-C (because there is less of it) and most of the UV-B radiation are prevented from reaching the ground.

Ozone is also sometimes broken up simply by colliding with a free oxygen atom. When this happens, the ozone molecule and the free oxygen atom combine to form two oxygen molecules. Ozone can also react with a variety of oxygen compounds and be transformed back into oxygen.

It is this very tendency of ozone to react with other molecules and turn back into oxygen that poses the problem that exists today. Before humans began polluting the atmosphere, a delicate cycle was in place in which oxygen was constantly being turned into ozone and ozone back into oxygen. This cycle can be disrupted, however. It is not a question of there being a limited supply of ozone that is used up, but one of upsetting the balance between oxygen and ozone. The introduction of too many artificial catalysts into the recipe can throw things off and change the relative concentrations of these two gases, potentially leading to a drastic alteration of the atmosphere.

The Concorde SST, or supersonic transport, is an impressive aircraft, but its nitrous oxide exhaust is an "ozone eater."

THE HUMAN FACTOR

The oxygen-ozone balance has continued for hundreds of millions of years, the last few million with human beings populating the planet. For a long time, no one thought that human activity could disrupt the awesome mechanisms that regulate planetary climate and atmospheric conditions. It all seemed too big, and human beings too small, for there to be any danger of mankind's upsetting the natural scheme of things.

In 1966, however, that danger became a stark reality. At that time, the United States, Britain, France, and the Soviet Union were all making plans to create fleets of SSTs, or supersonic transport planes, that would fly faster than the speed of sound. Excitement was boundless until Dr. John Hampson of the

Canadian Air Research Defense Establishment published a technical note pointing out that, unlike all other aircraft, the SSTs would fly through the lower part of the stratosphere, where the ozone layer lay. One of the principal exhausts of the SSTs would be water vapor, usually considered a harmless gas. But Hampson pointed out that ozone is so reactive that even water vapor would react with it, turning it back into oxygen. Could this mean, he asked, that more ultraviolet light would reach the earth?

By 1971, enough published reports on the dangers of SSTs to the atmosphere had reached the public that Congress called a hearing to determine whether the United States should continue the SST program. At the hearing, Dr. James McDonald, an atmospheric physicist at the University of Arizona, concluded that multinational fleets of SSTs could in fact result in a depletion of the ozone layer of several percent. A depletion of even 1%, he said, would cause 5,000 to 10,000 new cases of skin cancer every year in the United States alone.

At the same time, the Department of Commerce convened an SST advisory panel in Boulder, Colorado. Harold Johnston of the University of California at Berkeley was invited to attend because of his work involving the effects of nitrous oxides in the atmosphere. (Nitrous oxides are another major exhaust component of SSTs.) After much calculation he announced to the panel that a full fleet of SSTs would deplete the ozone layer by at least 10% in only 2 years. This would be disastrous for life on earth. As McDonald told the panel, "The whole history of evolution has been a battle with ultraviolet, and so far we've just barely won."

Representatives of the Boeing Aircraft Corporation, one of the companies hoping to build the SSTs, argued that all this was just a theory yet to be proved or possibly even a conspiracy against the industry and against progress. But before arguments could really get into full swing, the U.S. House of Representatives decided to cut off all funds for SST development. It appears that Congress shut down the SST program for economic reasons rather than environmental ones—it simply was not good business for the country to invest in the SST program at that time. One can only wonder whether, had it been profitable, the government would have approved the program despite the prospect of many thousands of new skin cancer cases yearly. At any rate, the United States did not develop an SST fleet, and the ozone layer was safe for the time being. The British-French Concorde SSTs now operating are too few to have a significant effect.

As most people now know, however, the abandonment of the SST program was hardly the end of what McDonald called the battle between humankind and ultraviolet radiation. For the first time, people were realizing that their own activities could have a significant impact on the earth and its atmosphere. And when British scientist James Lovelock found refrigerator and deodorant by-products up in the stratosphere, where, it turned out, they were harming the ozone shield, scientists began to realize that a major transformation of the planet might already be under way.

The hazards of such pleasures as beachgoing will probably escalate as the destruction of ozone allows more cancer-causing ultraviolet radiation to reach the earth.

chapter 2

CRISIS IN THE SKY

Most people are now aware that the earth's ozone layer is in danger. They also know that one of the major threats to this ozone has been aerosol spray cans. Many are even familiar with the scientific name for the dangerous ingredient of these aerosols: *chlorofluorocarbons,* or CFCs. But as recently as 1972, hardly anyone outside of the chemical industry had heard of CFCs, and spray cans seemed as harmless as chicken soup. Then James Lovelock had an interesting idea.

Lovelock, who 10 years earlier had been a tenured researcher at the British National Institute for Medical Research, was by 1970 an independent scientist working out of his home in Bowerchalke, in the southwest corner of England. This was by choice; Lovelock had decided that rather than do research for other people, he preferred to generate his own projects and then try to get the government or private companies to fund them. He had been successful enough working on his own to be paid a retainer by the Hewlett-Packard computer company and to have designed experiments that were carried on the *Viking* missions to Mars. He was also elected a Fellow of the Royal Society, Great Britain's prestigious scientific organization.

Although he was the first person to do research on CFCs in the atmosphere, Lovelock is best known today for developing the controversial Gaia hypothesis. This idea conceives of the earth as a sort of organism, called Gaia, that regulates its various physical systems in much the same way that the human body does. Just as the human body automatically regulates body temperature, fights against disease, and adapts to changing conditions, so too does Gaia maintain a stable global environment that sustains life, adjusting to various stimuli by means of natural feedback processes. The delicate oxygen-ozone balance would seem an apt illustration of this concept. In 1970, however, Gaia was a seed of thought years away from being born in Lovelock's brain. He just wanted to measure chloro-fluorocarbons.

Lovelock's reason for measuring CFCs was not that he suspected they were damaging the ozone layer. Quite the contrary; CFCs had always been considered completely harmless, unlikely to react with anything. Lovelock wanted to measure CFC concentrations essentially because he had developed a machine that could do it. The reason that he gave to potential research funders was that the compound might be used as a tracer to study atmospheric currents. Because most CFCs have been produced at known sites in the Northern Hemisphere, much could be learned about the circulation of the air around the globe by measuring the chemical compound's distribution over points in the Southern Hemisphere.

After a series of disappointments, Lovelock finally got some funding for his experiments, and in 1971 he sailed to Antarctica and used his new instrument to measure CFC concentrations there. His device was incredibly sensitive, detecting concentrations in units of parts per trillion. In early 1973, he published his findings in the respected scientific journal *Nature*. To everyone's great surprise, including his own, he had found that CFCs from industrial products were floating up into the stratosphere and accumulating there. Unfortunately, the discovery was considered nothing more than a curious finding, even to Lovelock himself. To his dismay even now, Lovelock concluded in his paper that "the presence of these compounds constitutes no conceivable hazard."

THE PERFECT PRODUCT

Chlorofluorocarbons were first discovered in 1928 by Thomas Midgley, an organic chemist at General Motors Corporation. Midgley was looking for inert (nonreactive), nontoxic, nonflammable compounds with low boiling points that could be used as refrigerants. Refrigerators work by allowing heat to flow into a liquid with a low boiling point. This liquid, called a coolant, absorbs the heat as it vaporizes, then is moved somewhere else and allowed to condense into a liquid again, releasing the heat. The heat is blown out of the system into the surrounding air, and the coolant remains to absorb more heat. Midgley found what he was looking for in the form of two compounds: dichlorodifluoromethane (CCl_2F_2, commonly known as CFC-12) and trichloromonofluoromethane ($CFCl_3$, called CFC-11). In each compound, different amounts of chlorine and fluorine are combined with methane, which is itself a combination of carbon and hydrogen—hence the name chlorofluorocarbons, or CFCs. These two CFCs were eventually manufactured by E. I. du Pont de Nemours & Company (known as

Du Pont) under the trade name Freon and by 1938 constituted 15% of the market for refrigerator gases.

CFCs seemed to be the perfect answer for cooling refrigerators and also air conditioners, which operate similarly. They were easily turned to liquid at room temperature with the application of just a small amount of pressure, and they could then easily be turned back into a gas. They were completely inert, reacting neither with other chemicals nor with surfaces, and were not poisonous to humans. For these same reasons, they were ideal as industrial solvents (for example, to clean intricate semi-conductor materials without ruining the plastic boards on which they were mounted) and as hospital sterilants. They were eventually also used to blow liquid plastic into various kinds of foams, from building insulation to fast-food hamburger containers.

Shortly after their invention as a refrigerant, another use for CFCs materialized. In the 1930s, household insecticides were dispensed with a long tube and plunger apparatus connected to a can of the poison. This bulky and inconvenient device provided the necessary reduction in pressure to vaporize the insecticide. CFCs, however, could easily be kept in a liquid form in an only slightly pressurized can. At the touch of a finger on the nozzle, the pressure could be lowered, enabling the CFCs to shoot out in their natural room temperature gaseous form, along with the insecticide itself. The spray can was born, and by 1947, 45 million of these CFC insecticide cans were sold each year.

It is important to distinguish between the terms *aerosol* and *spray can* because people have come to incorrectly use the two interchangeably. An aerosol is any fine mist of tiny droplets. Spray cans have been called aerosols because the fine mist they emit from their nozzles is an aerosol. What many people do not realize is that the word *aerosol* could just as easily refer to a cloud of water vapor droplets high above the earth as to an industrial tin can.

Insecticides were only the first application for CFC spray cans. They were soon employed for a multitude of products, from hair spray to deodorants. Finally, people could make themselves smell nice at the touch of a finger. They could also keep their hair in place and freshen the smell of their bathrooms, all at no apparent risk to their health or to the environment. The CFCs seemingly reacted with nothing and no one, disappearing harmlessly into thin air. If ever science and industry had combined for the benefit of humanity, this was it. In 1954, 188 million cans were sold in the United States alone. Four years later, that number increased to 500 million, and in 1968, 2.3 billion spray cans found their way into American homes.

In addition to chlorofluorocarbons, a number of related chemical compounds were developed over the years, all of which contained combinations of the five elements known as *halogens*: fluorine, chlorine, bromine, iodine, and astatine. These included bromine-based halons, used in fire extinguishers, trichloro-ethylene (later replaced by methyl chloroform), an industrial and dry-cleaning solvent; and carbon tetrachloride, used in fire extinguishers, grain fumigation, and CFC production.

Within a few decades, both the industrial and consumer worlds had come to rely on CFCs and other halogen compounds, under the misconception that these chemicals were harmless. Then Lovelock found that CFCs were sticking around in the stratosphere. And shortly after that, F. Sherwood ("Sherry") Rowland and Mario Molina decided to find out what effect the compound might be having up there.

In the fall of 1973, Molina, a postdoctoral chemistry student, came to the University of California campus at Irvine to work under Rowland, a respected scientist in the field of radiochemistry. Rowland was looking for something new to work on. He had become interested in CFCs after attending an atmospheric chemistry meeting in Florida the previous year, where he had spoken with Lester Machta of the National Oceanic and Atmospheric Administration (NOAA). Machta had been at a previous meeting that Lovelock had also attended, and they had spoken between lectures. The two had batted around some figures and had come to the interesting conclusion that the amounts of CFCs measured in the atmosphere were approximately equal to the sum total of all the CFCs that had been produced in the world up to that time.

Rowland was intrigued. If nothing in the troposphere broke apart CFCs, he thought, then they must eventually head up to the stratosphere. Once they got high enough, they would encounter high levels of ultraviolet radiation that would undoubtedly break them apart. Rowland remembers telling Machta, "Of course, it will always decompose with ultraviolet," but at the time it did not occur to him that the compound might pose a serious environmental threat.

Rowland suggested to Molina that it might be worthwhile for them to study exactly what happens to CFCs in the atmosphere. Molina's initial findings were interesting but unremarkable. He concluded that CFCs would not be affected in the troposphere but would in fact break down in the stratosphere under the onslaught of ultraviolet radiation, resulting in the release of free chlorine atoms. These chlorine atoms are very reactive—that is, they have a strong tendency to combine with other atoms or molecules. The ozone in the stratosphere is also very reactive, and so, Molina determined, each free chlorine atom would combine with one oxygen atom in an ozone molecule (O_3), leaving a chlorine-oxygen compound (ClO) and a molecule of oxygen (O_2). In this way, ozone would be broken down by the chlorine atoms from the CFCs. He decided, however, that CFCs would never be released in large enough quantities to have any noticeable effect on the ozone layer.

But as he continued to fiddle with the equations, Molina came to a startling new conclusion. When he suggested it to Rowland, they both realized they had an important discovery on their hands.

Chemists F. Sherwood Rowland (left) and Mario Molina, of the University of California, Irvine, were the first to realize the enormous threat posed by CFCs to the ozone layer.

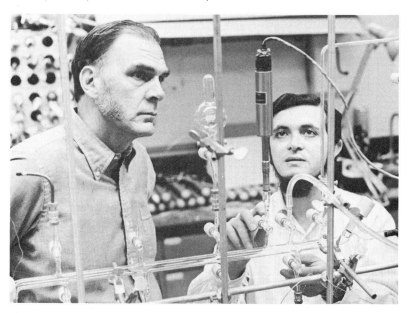

In fact, the formation of ClO and O_2 was not the end of the story. Not only do CFCs dissociate in the presence of ultraviolet radiation, but oxygen molecules also break down into free oxygen atoms. These free oxygen atoms then react with the ClO that is produced from the reaction between chlorine atoms and ozone, forming a molecule of oxygen and a free chlorine atom. The startling bottom line is that the original chlorine atoms, ozone molecules, and free oxygen atoms change into just chlorine atoms and oxygen molecules. The ozone turns into oxygen, and the chlorine that catalyzed the reaction remains intact, ready to react with more ozone. In other words, the chlorine is recycled over and over again.

Rowland and Molina calculated that each chlorine atom would turn approximately 100,000 ozone molecules into oxygen in this way before eventually being swallowed in some other, unrelated reaction (chlorine atoms can survive in the stratosphere for up to 100 years or longer). This meant that 100,000 times as much ozone would be destroyed as they had previously calculated. The two scientists now realized that the ozone layer was in very serious danger from the CFCs produced by human industry.

Rowland and Molina published their findings in the June 1974 issue of *Nature*. In this paper, their tone was one of controlled caution. "Important consequences may result," they wrote, and "more accurate estimates . . . need to be made . . . in order to ascertain the levels of possible onset of environmental problems." Despite this conservative overtone, their findings began to hit a nerve in the CFC industry. Du Pont, owner of the Freon trademark, was incensed to learn that its product had been connected with a completely theoretical danger to the environment. But it was not until September 1974 that the issue really hit the media.

The American Chemical Society (ACS), an enormous organization of chemists from industry, academia, and government, was holding its annual meeting that month in Atlantic City, New Jersey. The organization's news manager, Dorothy Smith, had the responsibility of picking a few of the more newsworthy papers to be read at the ACS conference and presenting them to the media. Sensing that a condemnation of Freon and other household CFCs was news, she picked the Rowland-Molina paper as one of the most interesting and, despite opposition from Du Pont, organized a press conference exclusively on the CFC-ozone issue.

Rowland's presentation at this event was more ominous than the *Nature* article. He and Molina now estimated that if CFC production continued to increase at the present rate of 10% a year until 1990 and then leveled out after that, the effect on the ozone layer would be a depletion of between 5% and 7% by

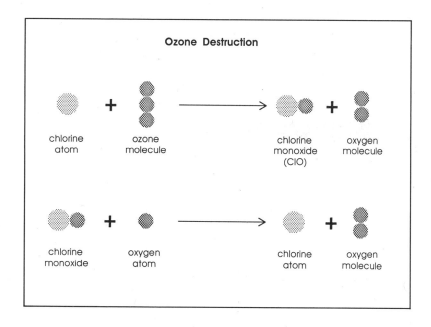

Ozone Destruction

chlorine atom + ozone molecule → chlorine monoxide (ClO) + oxygen molecule

chlorine monoxide + oxygen atom → chlorine atom + oxygen molecule

1995 and between 30% and 50% by the year 2050. To underscore the catastrophic implications of such a depletion, Rowland estimated that a depletion of 5% in the ozone layer would result in 40,000 extra cases of skin cancer each year in the United States alone. In the 1980s, the U.S. Environmental Protection Agency (EPA) estimated that an ozone depletion of 50% could cause more than 150 million cases of skin cancer worldwide by 2075. The majority of these would be basal- and squamous-cell cancers, which, although serious, are usually curable if treated in time, but increased UV-B could also lead to a rise in cases of melanoma skin cancer, which is usually fatal.

More ultraviolet radiation at ground level could also cause millions of cases of cataracts (a clouding of the eye lens that results in blurred vision) in the next century. Many medical experts also believe that a significant increase in UV-B would disrupt the human immune system, potentially leading to epidemics of infectious diseases. Other animal and plant organisms may suffer as well. Ultraviolet radiation has been shown to cause eye cancer in cattle. Studies also suggest that it may decrease crop yields of soybean, corn, rice, and wheat, as well as kill or damage aquatic plants such as phytoplankton, the base of the ocean food chain. Although much research is still necessary, it is clear that increased UV-B radiation will have widespread and—depending on the degree of its increase—possibly catastrophic effects.

Another potential effect of ozone depletion would be a disruption of weather on earth. Because ozone heats the stratosphere by absorbing solar radiation, it creates a natural temperature inversion over the cooler troposphere, inhibiting vertical air circulation. If the ozone layer is weakened, the atmosphere's circulation patterns could change, perhaps altering this planet's weather in as yet unknown ways.

An important aspect of predictions regarding the effects of ozone depletion is the future threat posed by CFCs already existing in the atmosphere. Because it takes many years for CFCs to travel from spray cans or air conditioners up through the troposphere to the stratosphere, there is a delayed effect in play. Even if CFC production were immediately halted—an event that will certainly not take place anytime soon—there would still be an increasing depletion of ozone for years to come due to the billions of tons of CFCs that have already been released into the air over the last 50 years. A study published in 1974 by a research group at Harvard found that "even if Freon use were terminated as early as 1990, it could leave a significant effect which might endure for several hundred years."

Various estimates of future ozone damage followed on the heels of Rowland and Molina's predictions; none of them were optimistic. The Harvard group predicted that at the present 10% rate of increase, the stratosphere would lose 10% of its ozone by 1994 and 40% by 2014. Another team at the University of Michigan reported that their computer predicted a 10% loss of ozone by between 1985 and 1990. The actual numbers may have varied, but there was a consensus: CFCs were destroying the ozone layer, and something had to be done about it.

On September 26, 1974, a headline on the front page of the *New York Times* read, TESTS SHOW AEROSOL GASES MAY POSE THREAT TO EARTH. The ozone war had begun.

The campaign against chlorofluorocarbons eventually led the McDonald's Corporation to stop using CFC-foam containers, which keep hamburgers warm but at a cost to the ozone layer.

T H E   F I R S T   O Z O N E   W A R

The mid-1970s were characterized by a battle between environmentalism and economics, between factions dedicated to saving the ozone layer at any cost and those reluctant to liquidate a thriving multibillion-dollar industry because of an unproven scientific concept.

In October 1974, the ozone problem was addressed on a governmental level for the first time. The National Academy of Sciences (NAS), a U.S. government–sponsored organization, appointed a five-person panel, including Rowland, to evaluate the situation. This group recommended an immediate one-year investigation to determine how serious a threat existed and whether enforced restrictions or bans on CFCs were in order.

In the meantime, the CFC industry was preparing to fight back. CFCs had been an enormously profitable business for 40 years, and they were not about to be abandoned simply because of 2 scientists' laboratory calculations. An industry group called the Manufacturing Chemists' Association funded a $5 million project to research the hazards of CFCs. Meanwhile, realizing the unlikelihood of winning in the laboratory, CFC proponents began a public relations counterattack. An editorial in the industry trade journal, *Soap/Cosmetics/Chemical Specialties*, protested, "The press, television, radio and other segments of the media seem to be carrying on a vendetta against pressure packaging. . . . Pressure packaging has served the world well and deserves a break." Robert Abplanalp, who had made millions of dollars by inventing the valve that made possible the lightweight aluminum spray can, went even further. "Extremists in the areas of ecology and consumer protection," he said, "are today waging a more effective war on American industry than the most capable host of enemy saboteurs."

Du Pont, the largest of the six U.S. manufacturers of CFCs, rallied behind its product in a more rational and professional manner. Company officials argued that it would be rash to dismantle an industry that provided 200,000 jobs and generated $8 billion in business annually for the sake of an unproven theory. "We welcome the scientific interest to develop hard, experimental facts about fluorocarbons and the atmosphere," a Du Pont official told the *Los Angeles Times*. "We believe that when this data is in hand, it will exonerate fluorocarbons."

In December, the House Subcommittee on Public Health and the Environment began the first government hearings on the ozone issue. Rowland testified and posed the question "How long should we wait for someone to find the missing factor . . . before we act?" Michael McElroy of Harvard warned that "if we wait too long the damage to our atmosphere may be so great that we'll have a difficult time repairing it." Industry spokespeople, meanwhile, continued to maintain that the theory was speculative and that regulations were still unwarranted. Unfortunately, Congress adjourned just days into the hearings, and nothing concrete was resolved. Still, the problem had been brought to national attention. It would be only the beginning.

In 1975, Senator Dale Bumpers of Arkansas chaired a small congressional committee called the Subcommittee on the Upper Atmosphere. Its main purpose was to determine whether the threat to the ozone layer was dire enough to warrant legislation restricting CFC use. Testimony was heard throughout the year from both sides, but although Bumpers maintained that "nobody has come before the committee and denied a single facet of the Rowland-Molina theory," no restrictive legislation was able to pass through Congress.

As it turned out, the first legislative response to the ozone crisis came at the level of state government. In May 1975, the Oregon legislature held hearings on the issue. Both Rowland and Molina flew up to Salem, Oregon, to testify, and their effort was not in vain. In June, Oregon became the first state to ban the sale of CFCs in spray cans, effective March 1, 1977. Echoing the cry of environmentalists, Governor Robert Straub said that even though all the facts were not in, he would rather "err on the side of caution."

The federal government, however, was still leaning toward caution in the other direction. In July, the Consumer Product Safety Commission turned down a petition to ban CFCs, stating that there was insufficient evidence that CFCs harmed the atmosphere. At the same time, the Department of Commerce released a study that concluded CFC regulations would impact 600,000 jobs with a payroll of $6.7 billion. Another 1.5 million workers, it said, depended indirectly on the CFC industry.

Of course, the ozone problem was not exclusive to the United States—all the developed countries of the world were using CFC products. Furthermore, upper-atmosphere currents distribute CFCs all over the planet; there are no political boundaries in the stratosphere. Still, the United States was by far the largest producer of CFCs, and it became the first political arena for the ozone war.

Throughout 1975, it looked as though the foes of CFCs were winning the war. While the debate raged on, with environmentalists urging foresight and preventive action and the CFC industry insisting that nothing had been proven, the public became convinced that the ozone layer was in fact being destroyed. For the first time, sales of CFC products dropped sharply.

In June, the Johnson Wax Company, the fifth largest manufacturer of spray cans in the country, announced an immediate self-imposed ban on CFCs. Within 60 days, their products would bear the label "Use with confidence, contains no Freon or other fluorocarbons claimed to harm the ozone layer." The rest of the industry was stunned. One official told *Rolling Stone* magazine, "What they've done is to try to gain marketing advantage out of a difficult situation. I know damn well that's what it is."

In early 1976, just as it seemed that the anti-CFC forces were winning the public relations battle, and just as the NAS panel was preparing a report that would conclusively back Rowland and Molina's predictions and probably lead to a ban on CFCs, a stumbling block appeared. Molina, hard at work back in Irvine to make sure that he and Rowland had overlooked nothing, came across some work done by German scientists involving a compound called chlorine nitrate. Rowland and Molina had known all along that the chlorine released from CFCs in the stratosphere could conceivably react with nitrogen compounds and form chlorine nitrate and that this reaction could tie up the chlorine and prevent it from destroying ozone. However, they had calculated that the chlorine nitrate would not stay intact long enough to have any important effect. Now, however, the German study showed that chlorine nitrate might be much more stable, throwing the Rowland-Molina calculations into doubt.

Mario Molina reentered the national spotlight in 1976 when he suggested that chlorine-nitrogen reactions might slow down CFCs' destruction of ozone.

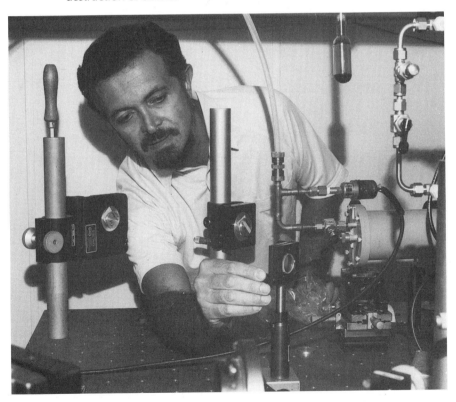

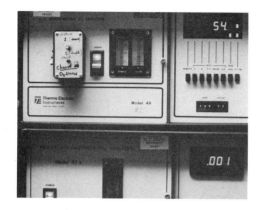

Instruments that measure concentrations of ozone and other gases help scientists determine what is happening to the earth's atmosphere.

Acting as a responsible scientist, Rowland phoned the news to the NAS panelists, who fed it into their computer model, a program that simulated climatic changes. The effect was shocking. Not only did the presence of chlorine nitrate seem to enormously decrease the depletion of ozone by chlorine, but it might even increase the ozone concentration. Instead of falling by 14%, the amount of ozone might actually rise by as much as 5%. Within weeks, these numbers changed as scientists found that the chlorine nitrate had less effect than originally anticipated, but in any case the fate of ozone was still up in the air.

Such fluctuations in experimental data occur often in science, but rarely under such public scrutiny. The CFC industry caught wind of these new uncertainties and wasted no time in using the news as ammunition. The media were right in the thick of things as well, and in the first months of 1976 the ozone issue made headlines all over the country. RUMOR AND CONFUSION FOLLOW OZONE-THEORY REVISION, read one headline. Another declared, AEROSOL SCARE MAY BE OVER. The public responded: After dropping by 14% the previous year, production of spray cans rose by 26% in February and by 41% in March. The CFC industry felt it was finally out of the woods.

That spring, the NAS committee, under enormous pressure to get out its report and befuddled by the confusing new evidence, summoned scientists involved in the controversy to an emergency meeting in Boulder, Colorado, at which they ran their computer models, determined to find out just what chlorine nitrate was doing to the chlorine-ozone reactions. It was weeks before the data compiled at this meeting were fully understood, but finally a consensus was reached. After all was said and done, the chlorine nitrate did not really have that much effect on the process. Most models now agreed that there would in fact be an ozone depletion, somewhere on the low end of Rowland and Molina's original 7% to 14% prediction. The "Aerosol Scare" was back.

After the delay imposed by the chlorine nitrate controversy, the NAS panel finally released its report in September 1976 at a packed press conference in Washington. The report found that Rowland and Molina were essentially correct. CFCs were damaging the ozone layer and should be restricted. If they continued to be released at 1973 rates, the ozone layer would eventually decrease by between 2% and 20%, with 7% the most probable amount. This depletion would cause an increase of 12% to 15% in ultraviolet radiation at ground level. Unfortunately for environmentalists, the report seemed to undermine its own findings in its second part, which dealt with the need for government regulation. It advised that the government wait two more years to study the problem before beginning to regulate CFC production.

The result was more confusion. SCIENTISTS BACK NEW AEROSOL CURBS TO PROTECT OZONE IN ATMOSPHERE, read a *New York Times* headline the next day, whereas the *Washington Post* announced, AEROSOL BAN OPPOSED BY SCIENCE UNIT. Rowland and Molina took the committee's findings as vindication of their research, but at the same time the CFC industry used the report as advertising material. "'We wish to recommend against a decision to regulate at this time,'" an industry representative quoted the panel's report in a newspaper ad, to which the advertiser happily added, "We agree!"

In the wake of this ambiguous report, proponents of CFC bans continued their efforts. Russell Peterson, chairman of the White House Council on Environmental Quality, said, "I believe firmly that we cannot afford to give chemicals the same con-stitutional rights that we enjoy under the law. Chemicals are not innocent until proven guilty. From the public-policy standpoint, there remains no valid reason to postpone the start of the regulatory procedures." Wilson Talley of the Environmental Protection Agency responded with frustration to the NAS call for two more years of study, saying, "We just cannot postpone decisions on these problems indefinitely in the hope that better data may be available in the indefinite future."

This point of view had considerable momentum, and there was nothing the CFC industry could do to stop it. Canada announced it was going to introduce immediate regulation of the industry. Later in the year, Sweden became the first European country to ban CFC propellants. On October 15, just one month after the NAS report had come out, the EPA and the U.S. Food and Drug Administration both proposed a phaseout of all non-essential uses of CFCs. The industry protested loudly, incredulous that the government could ignore the second, cautionary portion of the NAS report, but its opposition was in vain. On May 11, 1977, the EPA, the Food and Drug Administration, and the Consumer Product Safety Commission together announced a timetable for the elimination of nonessential CFCs. Manufacturing

Chlorofluorocarbons were invented for use as coolants in refrigerators because they are inert, or chemically stable, and have low vaporization temperatures.

would have to cease by October 15, 1978. Use of nonessential CFC products would end two months later. By April 15, 1979, interstate shipments would be banned.

The phaseout seemed to be a major victory in the struggle to save the ozone layer. The term *nonessential* applied to almost all spray can uses of CFCs. Fortunately for the chlorofluorocarbon industry, even while it had been lobbying against regulation of CFCs, it had been developing viable alternatives. Now, with the phaseout in effect, advertisements appeared touting new products that were CFC-free. Arrid Extra Dry deodorant cans, using non-CFC propellants, carried the slogan "Safe for the Ozone" in large letters on the side.

Despite the spray can victory in the United States, however, the ozone war continued worldwide into the 1980s. France and England remained opposed to CFC regulation, insisting that until actual ozone depletion was recorded, assumptions based on computer models were not enough to warrant action. Unfortunately, it was still impossible to determine with enough accuracy the actual extent of ozone depletion; by the time this technology was achieved, it might well be too late. So even though the U.S. cutbacks were enough to drop world levels of CFC production to pre-1973 rates, this improvement would be offset by increased use of the chemicals elsewhere in the world. Worldwide production was expected to grow at a rate of 7% annually from 1980 to 2000.

Another problem was that even as spray cans were being phased out, other uses of CFCs and related compounds were skyrocketing. As refrigerants, foam propellants, and in-dustrial solvents, they were still deemed "essential" and so were unaffected by the regulations. By 1983, worldwide CFC production was steadily increasing again, and people seemed to have assumed a false sense of security and confidence that the crisis was over. The ozone layer was no longer big news.

Then, in 1984, the public was once again shocked out of its complacency. Down in Antarctica, British researchers were discovering what came to be known as the "ozone hole."

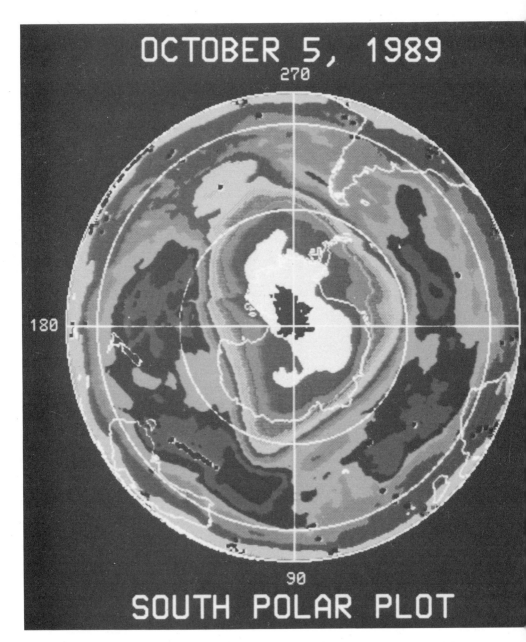

OCTOBER 5, 1989

270

180

90

SOUTH POLAR PLOT

A satellite image of the ozone hole over Antarctica, taken by Nimbus 7.

T H E   H O L E   A T   T H E
B O T T O M   O F   T H E   W O R L D

Researchers from the British Antarctic Survey have been measuring concentrations of ozone over Antarctica since 1957. Back then there was no apparent ozone crisis. The ozone measurements were just one facet of a worldwide scientific project called the International Geophysical Year, which aimed to learn everything possible about the global environment. The project was later extended to two years, and the British Antarctic Survey has continued to make ozone measurements every year since.

During the 1970s, the survey's leader, Joe Farman, and his colleagues found concentrations of ozone dropping slightly each September and October—springtime in the Southern Hemisphere. The percentage of decrease was small enough, however, to be within the margin of error of their aging spectrophotometer (a type of measuring instrument), and so they held off on publishing their results. It was still just an interesting trend.

In the polar spring of 1982, the British research team measured a depletion of about 20% of the ozone over the South

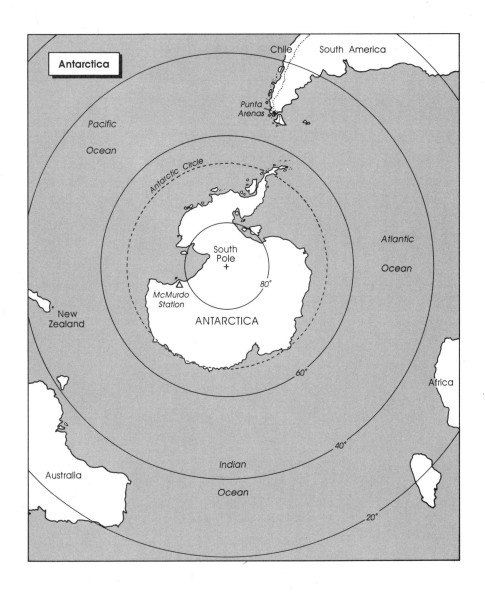

Pole. Normally, this would have been occasion for a sensational announcement to the science media. But in this case there was a reason for further caution. Since 1978, the American satellite

Nimbus 7 had been in orbit, carrying two major experiments designed to measure stratospheric ozone. These devices—the Total Ozone Mapping Spectrometer, or TOMS, and the Solar Backscatter Ultraviolet measurer, or SBUV—should have easily detected any major springtime ozone depletion. They were far more sophisticated devices than the British ground-based spectrophotometer and used an expensive computer system to analyze their data. But TOMS and SBUV found no unusual deviations at all in the annual ozone concentrations over Antarctica. The British kept silent and continued to take their measurements.

Then, in 1983, Farman's team began using a new, more precise spectrophotometer. Even as their margin of error shrank, the annual drop in ozone concentration over Antarctica seemed to be growing. In October of 1984 they could keep silent no longer. That month they measured a depletion of more than 30%, far more than instrument error could account for. Another measurement a thousand miles to the north corroborated the finding. *Nimbus 7* still reported nothing, but the British team was certain and sent off a paper with its results to *Nature*.

The paper, published on May 16, 1985, caused quite a stir in the scientific community. The most extreme reaction was probably at the NASA/Goddard Space Flight Center, the control center for the TOMS and SBUV experiments. Goddard scientists immediately rushed to their computers, and the problem soon became apparent.

The *Nimbus 7* experiments were designed so that their data were sent directly to the computers and processed before any humans were able to study them. This was part of the sophistication that supposedly made them superior to the old-fashioned techniques of the British Antarctic Survey. One of the nuances of the computer program, however, was a loop that got rid of any data that were "clearly" erroneous. That meant that any numbers outside of a wide range of reasonable possibility were automatically disregarded. This included the ozone concentrations recorded by Farman's spectrophotometer, which were so far below any that had ever been recorded that the *Nimbus* computer had simply ignored them. Luckily, it did not throw them out altogether but stored them in a backup file, so the people at Goddard were able to retrieve them and see what their system had missed.

With the "erroneous" data correctly processed by its computers, TOMS now told its scientists that an enormous hole in the ozone layer was indeed developing over Antarctica each southern springtime. The "hole" was actually more like a slice taken out of a thick layer, a slice as deep as Mount Everest is tall, covering an area as large as the continental United States.

It was shocking news. And now that the ozone hole was a confirmed fact, scientists set to work on another question: Why was this happening?

EXPLAINING THE HOLE

Although various types of solar or volcanic activity could have created the hole, the first culprit to investigate was chlorine. The British Antarctic Survey had been measuring CFC concentrations as well as ozone since Lovelock's work came to light in the early 1970s. Now they compared graphs of both ozone and CFC concentrations over that time period and found a direct correlation. Just as the concentration of CFCs rose, the concentration of ozone measured each October fell. This was not proof of a cause-and-effect relationship, but it was unlikely to be a coincidence. The next step was to figure out why conditions in the springtime might enhance chlorine-ozone reactions over the South Pole and thus create the hole.

By the end of 1986, it seemed clear that the answer had something to do with ice clouds. Several groups of researchers in both the United States and Germany independently came to similar conclusions regarding high-altitude clouds that form over Antarctica in the winter. These *polar stratospheric clouds*, or PSCs, form a diaphanous cloud covering in the lower strato-sphere, consisting of billions of tiny particles of ice. Scientists deduced that as these clouds form at the beginning of winter, CFC molecules become attached to the abundant surface area of the ice particles, triggering what are called *surface catalysis reactions.*

A thin layer of polar stratospheric clouds stretches across the sky 39,000 feet (nearly 12,000 meters) above Norway.

In the case of CFCs, the reaction that is catalyzed is the dissociation of the CFC molecules into free chlorine atoms. The chlorine atoms then remain attached to the ice particles throughout the long Antarctic winter. When rising springtime temperatures dissipate the clouds, the chlorine atoms are released into the stratosphere, where they wreak havoc on the ozone. When PSCs form again at the end of fall, the chlorine atoms are once again captured, and the ozone can recover through oxygen-ultraviolet reactions.

The reason that the hole was appearing only over the Antarctic is that it was the only place where stratospheric temperatures dipped low enough to create PSCs. In fact, worldwide ozone depletion may have led to a recorded drop in October temperatures in the Antarctic stratosphere of 32.5 degrees Fahrenheit (18 degrees Celsius) between 1979 and 1985. This probably occurred because ozone helps to heat the stratosphere by absorbing ultraviolet light. There may be a positive-feedback mechanism at work whereby decreased ozone leads to colder temperatures that create more PSCs, which in turn help destroy even more ozone.

ANTARCTIC EXPEDITIONS

From August through October of 1986, the first National Ozone Expedition (NOZE) took place. Scientists traveled to McMurdo Station in Antarctica to study the ozone hole with ground instruments and experiments loaded on weather balloons, assisted by the TOMS and SBUV experiments on *Nimbus 7*. Balloons launched from McMurdo Station recorded startling depletions of ozone between 7.5 and 12.5 miles (12 and 20 km) above the earth; in that region about 35% of the ozone disappeared between August and October. In a smaller region between 8 and 11.5 miles in altitude, more than 70% of the ozone measured in August had disappeared. One reading in October measured an incredible loss of 90% of the ozone in a region between 1 and 3 miles thick—an almost complete hole in the ozone layer.

The NOZE experiments also found large concentrations of chlorine compounds at just the altitudes and times when the ozone was disappearing. There was now little doubt that chlorine from CFCs, with the help of polar stratospheric clouds, was creating the hole in the sky.

The next major ozone experiment came the following year, in the Antarctic spring of 1987. In August of that year, scientists from the United States, Great Britain, Argentina, Chile, and France converged on the city of Punta Arenas, Chile, for the Airborne Antarctic Ozone Experiment (AAOE). Punta Arenas is the southernmost city in the world and is best known as a staging area for scientific flights to the Antarctic. Organized by the National Aeronautics and Space Administration (NASA) along with the National Science Foundation, the National Oceanographic and Atmospheric Administration, and the U.S. Chemical Manufacturers Association, the AAOE consisted of 21 experiments mounted on 2 aircraft. The first of these aircraft was a converted DC-8 that carried an entire team of scientists along with their apparatus and flew at an altitude of 7.5 miles, just at the bottom of the ozone layer. The other plane, which captured much more public attention, was the ER-2, a scientific version of the famous U-2 spy plane of the 1950s. The ER-2 flew at an altitude of 12.5 miles, virtually as high as is possible without a

Ross Island, as seen from McMurdo Station, Antarctica, the staging point for the 1986 and 1987 National Ozone Expeditions.

rocket-powered spaceship, and had room for only a single human being—the pilot. The scientific instruments were packaged into pods, or containers, located under each wing, and were operated by the pilot with remote hand controls.

The ER-2 had the advantage of flying right into the heart of the ozone layer, and on its 12 voyages, totaling more than 40,000 miles (64,000 km), it measured temperature, ozone concentrations, aerosol particles, water, chlorine compounds, nitrogen compounds, and other ozone-affecting compounds in the vicinity of the aircraft. The DC-8, although it flew lower, had two advantages: It had a much greater range, allowing for sorties farther over Antarctica; on its last voyage it actually flew completely across the Antarctic and landed in New Zealand. It also had instruments similar to the spectrophotometer that allowed it to take measurements in a column of atmosphere stretching six miles above the aircraft.

Other research was being done simultaneously with the AAOE. Back at the Goddard Space Flight Center, scientists processed data from TOMS and SBUV in order to complement the AAOE work. Ozone maps from the TOMS computer were immediately sent to the Puntas Arenas control center so that they would know where the best spots to measure ozone depletion were located. In addition, NOZE II, a follow-up to the previous year's expedition, was working out of McMurdo Station in coordination with the Punta Arenas group. Related work was conducted from stations at Halley Bay and the South Pole.

Normally, it would take many months to analyze data of this nature and more months before any results would be announced to the public. But because of the extraordinary circumstances surrounding these experiments—that quick regulatory

A balloon is launched from McMurdo Station to collect data on ozone depletion. Instruments are frequently carried to high altitudes in this manner in order to study atmospheric conditions.

legislation would be necessary in order to counter the trends that were being discovered and that the very fate of the earth was in question—ordinary time frames were compressed or dispensed with altogether. On September 30, 1987, just days after the completion of the final experiments, a press conference was held to announce the most obvious and crucial findings.

Robert Watson, the director of the AAOE mission, reported that after 6 weeks and more than 110,000 miles of flying, the researchers had found a definite link between chlorine and the annual ozone hole. They also found that virtually every compound that could have an adverse effect on ozone was present in abnormally high proportions during the springtime depletion, whereas substances that could boost the amount of ozone existed in abnormally low proportions. The ER-2 pilots reported flying through extensive PSC clouds that stretched across the sky in even greater abundance than had been anticipated. As was expected, stratospheric temperatures were much lower than they should have been had enough ozone been present to absorb ultraviolet light and radiate heat.

Other, ground-based experiments showed that ozone depletion in the spring of 1987 was worse than ever. Measurements between latitudes of 70° and 80° south showed 15% less ozone present than in 1985, which until then had been the greatest annual depletion. More than half the ozone was gone over this region. At Halley Bay, 76° south, balloon flights indicated that on October 15, 97.5% of the ozone that had been measured at 10.3 miles altitude on August 15 was gone.

Never before had a major scientific experiment's data been studied, processed, and revealed to the public so quickly. The Airborne Antarctic Ozone Experiment also provided what may have been the most shocking and ominous results of any scientific expedition in history—and the entire project took only three months. Now it was time for the ozone drama to move on to a new stage, in Berlin.

The occasion was the Dahlem workshop, held in the first week of November 1987. Most of the people involved in the work on the ozone problem had converged on Berlin for an informal brainstorming session to analyze the AAOE and other data. The participants included Rowland and Lovelock, the ozone pioneers of the 1970s, as well as Farman of the British Antarctic

Survey and Watson from the AAOE. By the end of the week, the scientists at the Dahlem meeting had reached several major conclusions.

The simplest and most urgent consensus was that chlorine from industrial products—in particular, CFCs—was definitely responsible for the disappearance of ozone over Antarctica each spring. Watson presented to the group a graph of measurements taken from the ER-2 flights through the ozone hole. As the plane passed through the area of ozone depletion, the amount of chlorine compounds rose just as the ozone concentration fell. He then showed a magnification of the graph, showing every nuance of the curves, and everyone could see that for each minute rise or fall of ozone, there was a corresponding fall or rise in chlorine. The relationship was too exact to be coincidental. As science writer John Gribbin remarked in his book *The Hole in the Sky*, "People talk of finding the 'smoking gun' that incriminates CFCs; this [graph] is more in the nature of a signed and witnessed confession."

The information presented in Berlin confirmed the general theory behind the ozone hole. CFCs become attached to ice particles in the high stratospheric clouds, or PSCs, that form at the beginning of each winter. There they are dissociated into free chlorine atoms and stored through the long polar winter. When the sun reappears in spring, the chlorine atoms are released and begin to destroy the ozone. Come winter, the chlorine is tied up in the PSCs once again, and ozone has the chance to re-form.

As long as PSCs form only in the Antarctic, the scientists figured, this would probably be the only place that the "hole" effect would occur. Even near the North Pole, temperatures did not drop low enough to form PSCs. But the danger remained that conditions that led to more abundant PSCs in Antarctica could spread to other areas of the world. PSCs may have always existed down there, but almost certainly not in as great an abundance as they do now. This is because the Antarctic stratosphere has become much colder in recent years, for two main reasons: First, as the ozone layer is depleted, it cannot absorb as much ultraviolet light, and so the temperature of the stratosphere plummets. This trend has probably already begun elsewhere in the world besides over Antarctica. The other reason is that gases in the upper troposphere may be trapping more heat beneath them than they used to, contributing to the lower temperatures above them, in the stratosphere. This latter concept is one that must be addressed when discussing recent atmospheric trends. It is called the *greenhouse effect.*

Among the potential consequences of global warming would be the melting of polar ice, which would lead to further temperature increases, disruptions in weather patterns, and rising sea levels.

O Z O N E   I N   A
G R E E N H O U S E

Although the greenhouse effect has become a familiar household phrase only in the last decade or two, it is not an entirely new concept. In 1896, the Swedish chemist Svante August Arrhenius took note of French mathemetician Jean Baptiste Fourier's observation that carbon dioxide in the atmosphere may act much like a greenhouse. Arrhenius wrote that if people continued to burn carbon-based fossil fuels, creating carbon dioxide as a by-product, they might actually raise the temperature of the planet.

This concept was touched upon in Chapter 1. Venus, with a carbon dioxide–rich atmosphere, is far hotter than can be explained by its nearness to the sun. And Mars, with a very thin atmosphere, is much colder than one would expect. Perhaps an even better example is the earth's moon, which is virtually the same distance from the sun as the earth, yet has an average temperature of about –13°F (–25°C), or 72°F (22°C) colder than that of the earth. The reason for this difference is that the moon has no atmosphere with which to trap heat—that is to say, it has no greenhouse effect.

The atmosphere is heated by the sun, but not in exactly the way many people think. It might seem logical that the atmosphere is heated directly by the sun's rays, but that is not the case. If it were true, then the higher one got, or the closer to the sun, the hotter it would be. But as anyone who has been in the mountains knows, temperature drops with an increase in altitude. This occurs because the atmosphere is actually heated by the earth.

As mentioned earlier, most of the radiation the earth receives from the sun is in the form of visible light, between 400 and 750 nm. This light passes through the atmosphere undisturbed; if it were absorbed or reflected, people on the ground would never see or feel it. In fact, much of the sunlight that reaches the earth is reflected, mostly by the enormous sheets of ice at the North and South poles. This property, called reflectivity, or *albedo*, plays a large part in determining the earth's temperature. The earth absorbs the light that is not reflected away, heats up as a result, and radiates it back into the atmosphere as infrared radiation. Infrared light has a much longer wavelength—between 4,000 and 100,000 nm—than visible light, and much of it is absorbed by water vapor and carbon dioxide in the atmosphere. When the air absorbs this infrared radiation, it heats up, "trapping" the sun's energy in the troposphere. The process gets its name because it works much like a greenhouse, which lets sunlight in and then traps the resulting heat inside its glass. Without the greenhouse effect, life could never have evolved on earth.

When Arrhenius sounded his warning a hundred years ago, it created no panic, nothing more than minor curiosity within the scientific community. People still believed that human activity was too small to have a noticeable effect on the enormous workings of the planet. This conviction was to last well into the 1960s, when both the ozone and carbon dioxide issues began to creep into the public consciousness.

THE FIRST GREENHOUSE DATA

In the late 1950s, scientists began to measure carbon dioxide concentrations in the atmosphere. They chose distant, isolated locations: on top of the dormant Hawaiian volcano Mauna Loa and within the Arctic and Antarctic circles. Their goal was to get readings undisturbed by the presence of local industrial effects, and their findings were alarming: They measured average concentrations of 315 ppm (parts per million, meaning that one liter of air contained 315 millionths of a liter of carbon dioxide). Studies of carbon deposits in tree rings show that in 1850, before the Industrial Revolution had caused widespread burning of fossil fuels, carbon dioxide concentrations in the air were only about 270 ppm. Before then concentrations had remained virtually constant in that range; since then they have increased drastically, today measuring about 350 ppm.

Scientists estimated that approximately half of the carbon dioxide currently produced by burning fossil fuels remains in the atmosphere. The rest disappears into *sinks*—natural processes that take carbon dioxide out of the atmosphere, such as plant photosynthesis and dissolving of the gas in oceans. Researchers then fed this information into their various computer models. As with the CFC-ozone problem, there was a wide disparity between computer predictions. By 1975, however, the different models had been analyzed and a consensus reached. Scientists generally agreed that a doubling of the carbon dioxide concentration from the original reading of 270 ppm would result in an average increase in global temperature of 3.5°F (1.9°C). At the present rate of production, that doubling could be reached as early as the beginning of the next century.

An increase of 3.5°F may not sound like much, but in fact it is a disturbance of catastrophic proportions. On any given day, a swing of a few degrees is inconsequential, but a constant *average* increase of that much over a period of centuries or even less is a different matter. To understand what a change of that magnitude can mean, consider that at the height of the last ice age, about 18,000 years ago, the average global temperature was just 9°F (5°C) colder than it is today, and yet all of Canada and much of the United States, down to about the latitude of New York City, was covered with glaciers more than a mile thick.

No one knows exactly what would happen if global temperatures rose by 3.5°F, but the best guesses are not optimistic. Warming could cause glaciers at the poles to melt. These glaciers are crucial because they reflect an enormous amount of the sun's light. If they began to melt, more light would be absorbed rather than reflected by the earth, and global temperatures would rise even more rapidly. This accelerated warming could have disastrous effects on weather patterns and agriculture and drastically alter coastlines. This has happened before in the history of the earth: About 100 million years ago, when the dinosaurs thrived, temperatures were about 4.5°F (2.5°C) warmer than they are today, and parts of Canada were as tropical as Miami Beach. Now global warming threatens to happen faster than ever before because of the production of

A temperature rise of just a few degrees could produce severe drought conditions in the northern United States and Canada, among other places.

greenhouse gases. How quickly could Canada adapt its agriculture and economy to a tropical setting?

Although the public remained largely unalarmed even after a 1975 article in *Science* predicted the 3.5°F rise, scientists were quite concerned. Almost all of the predictions were pessimistic. Although it is difficult to predict exact local conditions in a global warming, most scientists suggest that parts of the northern United States and Canada would suffer hotter, drier conditions, resulting in a huge dust bowl, while other parts of the continent would be ravaged by great new storm systems because the warming would cause increased evaporation of water, forming more clouds and thus creating much more "weather." These frequent storms would probably also occur over desert areas, but this would not be as beneficial as it may sound. A desert is a fragile environment, with no tree roots to hold it together, as in a forest. Great storms would simply wash away the land instead of nourishing it. In a worst-case scenario, in which the glaciers of west Greenland and Antarctica were to melt completely, many of the world's great cities, built intentionally on the water, would be submerged. Actually, these cities could probably be protected by massive construction projects, but low-lying areas of the Third World would not be, and millions of refugees could one day pour out of these areas. Most of civilization has been built around local climatic patterns; global warming threatens to disrupt these patterns and with them the prosperity of the human race.

OTHER GREENHOUSE GASES

So far, carbon dioxide has been treated as the sole culprit in the acceleration of the greenhouse effect. Other gases are responsible as well, however, and may together contribute as much to the effect as does carbon dioxide. Right now the main contributor other than CO_2 is methane, which is presently trapping about 36% as much infrared heat as carbon dioxide.

Methane, commonly called "natural gas," was around long before humans appeared on the earth; however, it existed in the atmosphere in a constant amount until about 300 or 400 years ago. At that point the amount of methane in the air began to increase dramatically. One cause is termites. These insects house inside their digestive system an ancient species of bacteria that generate energy by turning carbon into methane, which the termites then spew into the air. The effect is considerable; enormous pockets of methane have been measured around termite nests. Humans have contributed to this process by providing food for termites in the form of dead trees. Termites much prefer dead wood to live trees, and so as people have destroyed forest after forest around the world, the termite population has boomed, along with the methane-producing bacteria population. Termites now pump 75 to 200 million tons of methane into the atmosphere each year. Other major sources of methane include rice paddies, garbage landfills, and the stomachs of cows. It is also likely that global warming itself increases methane concentrations by speeding up bacterial breakdown of organic wastes.

The results are significant. The amount of atmospheric methane in prehistoric times, determined by analyses of Greenland and Antarctic glacial ice, was about 0.7 ppm. The

A street in Dhaka, Bangladesh. This country, which borders on the Bay of Bengal in southern Asia, is one of many low-lying regions of the world that could be flooded if global warming leads to higher sea levels.

current concentration is about 1.7 ppm and is rising at the disturbing rate of 1% or 2% each year. The amount of methane doubled once over the last 200 years; at the present rate, it will double again in just 70 years. It could very well soon overtake carbon dioxide as the number-one greenhouse gas.

Nitrous oxide is another potent greenhouse gas and one that also destroys ozone. This compound, commonly called "laughing gas" when used in anesthesia, is emitted in huge quantities by nitrate fertilizers as well as by fossil fuel combustion and the production of nylon textiles and plastics. Limited amounts of nitrous oxide produced naturally help maintain the oxygen-ozone balance. Emissions of the gas are increasing, however, and in excess it may pose a serious threat to the ozone layer. As a greenhouse gas, nitrous oxide is more than 200 times as powerful as carbon dioxide.

Another significant contributor to the greenhouse effect is a familiar word: CFCs. In 1973, James Lovelock was the first person to point out an additional danger of CFCs, aside from their ozone-destroying properties. At a conference on fluorocarbons, he suggested that if CFC concentrations in the atmosphere increased tenfold—which they now have—they could become dangerous greenhouse gases. Not only do CFCs react with ozone; they also have the property of absorbing infrared heat, just like carbon dioxide and methane. Lovelock suggested at the time that this greenhouse property of CFCs could eventually far outweigh the dangers they pose for the ozone layer. The National Academy of Sciences report on CFCs in 1976 included a study of their role as greenhouse gases, noting that even at an increase of only a few percent a year, CFCs could lead to a greenhouse effect of "drastic proportions." Already it is estimated that CFCs contribute to the

Cows are a source of methane, one of the more potent "greenhouse gases" that trap infrared radiation in the troposphere.

greenhouse effect almost 40% as much as carbon dioxide. Moreover, because CFCs released years ago are still making their way up through the troposphere, their contribution to global warming will probably continue to increase.

THE OZONE-GREENHOUSE CONNECTION

That CFCs act as greenhouse agents is not merely a curious coincidence; ozone depletion and the greenhouse effect are in fact intimately connected aspects of a changing atmosphere. Moreover, their effects may aggravate one another. For one thing, ozone is itself a greenhouse gas when it appears in the troposphere, which it does in greater amounts now due to depletions in the stratosphere. As mentioned in Chapter 4, where there is less ozone in the stratosphere, more ultraviolet light reaches the troposphere, where it can change oxygen into ozone. In addition to absorbing ultraviolet light, ozone also absorbs infrared radiation and so contributes to the trapping of heat in the troposphere.

Both ozone depletion and global warming also reinforce one another's effect on temperature in the troposphere and stratosphere. In other words, they both heat up the troposphere and cool down the stratosphere. Less ozone in the stratosphere causes less ultraviolet light to be absorbed there; temperatures drop as a result, because ultraviolet absorption is the main source of stratospheric warmth. Rowland estimated in 1987 that by the middle of the next century, ozone depletion will have produced a stratospheric temperature drop of about 36°F (20°C). Using a similar climate model, he predicted that global warming during the same time period would create a drop of 18°F (10°C) in the stratosphere while heating up the troposphere by 5.5°F (3°C). This phenomenon would occur because the greenhouse effect traps in the troposphere infrared heat that would otherwise rise into the stratosphere and be somewhat absorbed there. Ozone depletion and the greenhouse effect thus work together to cool off the stratosphere; by Rowland's prediction, the combined result of ozone depletion and the greenhouse effect 60 years from now will be a stratosphere 54°F (30°C) colder than it is now.

It is impossible to predict exactly how such a temperature change would affect weather patterns on earth, but such a significant alteration would undoubtedly create some problems. One possible consequence of this stratospheric cooling involves the polar stratospheric clouds that appear to cause the ozone hole over Antarctica each spring. Given a 54°F drop in stratospheric temperatures around the globe, it is not inconceivable that PSCs could form elsewhere, and it is possible that wherever PSCs form, the same surface catalysis reactions would lead to similar holes in the ozone layer. It may even be that global warming has already fomented the processes by which the Antarctic hole has formed, although temperature changes have not nearly approached the proportions Rowland predicts for the next century.

This PSC threat is just one example of how ozone depletion and the greenhouse effect can feed on one another and create a snowball effect that could lead to disastrous consequences. It is also possible that the various changes could work against each other—that is, the effects could offset each other. One possibility is that the cooler temperatures in the stratosphere might slow down the rates of the chemical reactions by which ozone is destroyed, as these reactions run faster in warmer environments. This argument is given in the same hopeful spirit as the suggestion that warmer temperatures on earth could lead to a more comfortable and nourishing world, with gentler winters and fewer damaging frosts.

The general feeling among the scientific community, however, is that these atmospheric changes would be traumatic for the environment rather than beneficial. The processes by which the atmosphere creates ozone, absorbs radiation, heats itself, and maintains relatively constant temperatures on earth are clearly fragile. Temperature swings of just a few degrees throughout the history of the earth have changed continents such as North America from enormous sheets of ice into tropical marshlands and back again. The present state of the environment, in which humankind has prospered, is just one phase in this continuing cycle. If people wish to maintain that environment, they had better not take their chances on what kind of world might be created by climate changes and instead focus on what they can do to prevent those changes.

At a Washington, D.C., news conference in June 1990, attorney David Doniger of the Natural Resources Defense Council displayed some of the 141 common spray can products that contain the ozone-destroying compound methyl chloroform.

chapter 6

THE STATE OF THE OZONE

It has been nearly 20 years since Lovelock first found CFCs in the stratosphere and Rowland and Molina discovered that the chemical compound was destroying ozone. One might think that upon hearing of the impending destruction of the vital ozone layer, the world community would waste no time taking steps to eliminate the causes. This is not the case, however. Despite local, national, and international actions taken against them, CFCs are still being produced around the world. And even as CFCs produced years ago are reaching the stratosphere and beginning to break down ozone, these new emissions are beginning their slow journey upward, where their effects will be felt far into the future.

Even now, ozone depletions have been recorded all over the globe, although not nearly at the levels found in the Antarctic hole. In 1988, NASA's Ozone Trends Panel found significant ozone depletion between latitudes 30° north and 64° north—an area that includes the United States, Canada, Western Europe, China, Japan, and the Soviet Union. Scientists recorded depletions of 1.7% to 3% over the entire year, with as high as 6% depletions recorded during the winter. This accelerated decrease in wintertime may be due to the same processes that create the Antarctic hole at the end of each winter.

It is also entirely possible that the opening of the Antarctic hole has already affected ozone concentrations across the Southern Hemisphere. Atmospheric currents are complex, but there are probably large sections of ozone-poor air leaving the Antarctic each springtime and blowing across the Southern Hemisphere. Stratospheric circulation evens out concentrations of gases, and the result is a dilution of the ozone. Jerry Mahlman of Princeton University has predicted that this effect could reduce the overall ozone concentration over the Southern Hemisphere by an additional 3% to 4% if the springtime Antarctic hole continues. To put that percentage in perspective, it is about the same amount of ozone that recent international legislation has aimed to save.

A HOLE IN THE NORTH

Perhaps the most significant development in the last few years has been the appearance of an ozone hole over the Arctic. It was previously thought that because temperatures over the North Pole are about 18°F (10°C) higher than over the South Pole, a hole similar to the one over Antarctica would not develop; temperatures would not be cold enough for the formation of the polar stratospheric clouds that catalyze the release of ozone-destroying chlorine. Now, however, PSCs have been found over the Arctic along with a hole in the ozone similar to, although much smaller than, the Antarctic hole.

In the early spring of 1989, a multinational group of scientists headed up to the Arctic Circle to spend 39 days studying the atmosphere over the area. The Airborne Arctic Stratospheric Expedition (AASE) found that at an altitude of about 12.5 miles, 15% to 20% of the ozone had been lost. This amount was far below the 50% to 95% lost at certain altitudes in the Antarctic, and total ozone destruction at all altitudes in the Arctic was only a few percent, but it was still an alarming discovery. Moreover, the researchers found large amounts of chlorine, almost certainly from CFCs, in the very areas where the ozone was most depleted.

Scientists are not certain of the reasons for the Arctic hole, but the most likely cause is the presence of PSCs. It is also not certain why PSCs are forming in the Arctic, but one strong possibility is lower temperatures caused by previous ozone depletion. As ozone decreases in general, less ultraviolet light is absorbed in the stratosphere and the temperature drops accordingly, leading to the formation of PSCs, which in turn catalyze the production of free chlorine, which destroys more ozone. This so-called positive-feedback process could very well lead to a larger hole in the Arctic, one on the same scale as that over Antarctica.

The findings of the AASE in fact showed that in 1989 the Arctic avoided an Antarctic-like gaping hole by the narrowest of margins. A hole did not occur for two reasons: First, nitrogen-rich air poured in from lower altitudes, which tied up chlorine in nitrogen compounds. Also, temperatures were slightly too warm. The AASE scientists warned, however, that these preventive conditions may not exist in future springtimes. The expedition report concluded that observations "clearly show that the Arctic stratosphere is primed for an ozone hole." As writer John Gribbin observed in April 1990, "All the chemical precursors of [an ozone hole like that in Antarctica] are in place. They are waiting only for the final trigger of a prolonged period of cold stratospheric temperatures, with little mixing of air from lower latitudes."

Another consequence that must be studied is the possibility that the Arctic hole is responsible for exporting ozone-poor air to lower latitudes. This could already be responsible for the depressed ozone readings across the Northern Hemisphere in the last few years, in much the same way that the Antarctic hole is probably affecting southern ozone concentrations. Dangerous chlorine created in the PSCs and even the PSCs themselves could be blown to lower latitudes by unpredictable Arctic wind patterns. These are just some of the questions the ozone scientists will be tackling in the future.

THE SECOND OZONE WAR

At the end of the 1970s, when the United States decided to phase out all "nonessential" CFC production, it seemed as if the ozone war were virtually over. But as so-called essential uses of CFCs, such as refrigerants and industrial solvents, continued to proliferate, total CFC production actually rose in the early 1980s. That growth, together with the discovery of the Antarctic hole in 1984, meant a continued battle throughout the 1980s between those pushing for elimination of CFCs right away and those insisting on a gradual phaseout. No one was arguing anymore that CFCs are not harmful; it was a question of how serious was the potential harm and whether it outweighed the cost of liquidating an enormous industry.

In 1980, the United Nations Environmental Program (UNEP) called upon the governments of the world to cut down on

A converted DC-8, NASA's "flying research laboratory" is shown on takeoff from Stavanger, Norway, during the Airborne Arctic Stratospheric Expedition of 1989. The aircraft can fly to an altitude of 40,000 feet.

their production and use of CFCs. The request was largely ignored, but in 1981, UNEP set up a committee to draft an international agreement to protect the ozone layer. It took several years, but in March of 1985, 20 countries met in Vienna and signed what came to be known as the Vienna Convention. The document was more a statement of intentions and principles than a blueprint for specific protective action. Its 21 articles declared the necessity for nations to cooperate with each other to save the ozone layer. It set up a basis for exchanging research findings and coordinating ozone monitoring with the understanding that more specific agreements would come in the near future. Although the convention spoke mainly in generalities, two technical annexes did spell out specific research that would need to be carried out in the immediate future. The treaty also got some immediate results: In 1986, the Soviet Union released figures on CFC production for the first time ever.

In 1986 and 1987, international debate raged over who should be doing the most cutting of CFC use and how much should be cut. The United States accused Great Britain and France of holding back the rest of the international community; Richard Benedick, a deputy assistant secretary for the environment, criticized these two countries for being "more interested in short-term profits than in the protection of the environment for future generations." On the other hand, the United States, specifically Du Pont, was far ahead of the rest of the world in the development of CFC substitutes. For this reason the U.S. government and industry were accused of pushing for a CFC ban so that they could dominate the substitutes market. In the meantime, countries still just beginning to develop their CFC industries, such as China and India, refused to take part in talks at all.

Despite all the problems, representatives from 43 countries managed to convene in Montreal in September 1987 under the auspices of UNEP to try to put together a concrete international protocol for ozone protection. For two days and most of one night, negotiators haggled over the various details and nuances involved in the making of an international treaty. The biggest stumbling block was a last-minute demand by the U.S. delegation that the treaty not go into effect until it had been ratified by countries whose total combined CFC consumption was 90% of the world's. This stipulation was made so that the United States would not have to begin their CFC cuts before any of the other major producers. It threw an enormous wrench into the negotiating process, however, because it would have meant that countries such as the Soviet Union, with 10% of world CFC production, or Japan, with 11%, could effectively veto the treaty.

Working practically through the night, negotiators finally came up with compromises that suited everyone. The United States backed down on its 90% demand and agreed to a

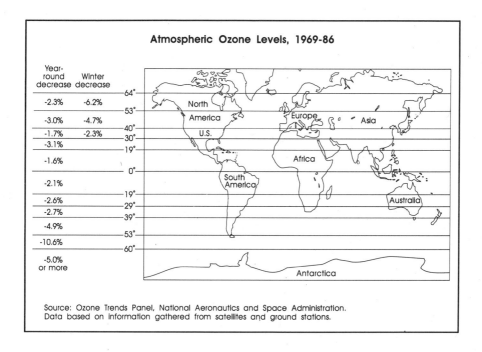

Atmospheric Ozone Levels, 1969-86

Year-round decrease	Winter decrease
-2.3%	-6.2%
-3.0%	-4.7%
-1.7%	-2.3%
-3.1%	
-1.6%	
-2.1%	
-2.6%	
-2.7%	
-4.9%	
-10.6%	
-5.0% or more	

Source: Ozone Trends Panel, National Aeronautics and Space Administration.
Data based on information gathered from satellites and ground stations.

stipulation that governments with two-thirds of worldwide CFC production be required to ratify the treaty before it went into effect. The Montreal Protocol, signed by all 43 countries on September 15, 1987, and effective as of January 1, 1989, called for the following measures: a freeze on consumption and production of all CFCs at 1986 levels by 1990; a cutback of 20% by January 1, 1994; and an additional cutback of 30%, for a total reduction of 50%, by January 1, 1999.

The Montreal Protocol was not nearly as potent as it might have been. The European contingent had kept the cutbacks down to 50% in order to postpone the dismantling of their CFC industries. Third World countries, such as India and Brazil, received extensions on the deadlines, giving them an additional

10-year period for more CFC growth. Even the Soviet Union got an exemption for certain CFC plants already planned or under construction. The treaty also did not restrict uses of most related halogen compounds, whose production was on the increase. Still, despite its weaknesses, the agreement in Montreal was an important step forward. Perhaps the most significant component of the treaty was the stipulation that the countries meet again periodically to review scientific evidence and with the intent to eventually eliminate all ozone-destroying products. Rowland pointed to this last element as the saving grace of a somewhat anemic if well-intentioned document. "I think the situation is more dangerous than most of the nations are acknowledging," he said. "I'm sure there will be a call for more [reductions] as soon as the ink is dry on this treaty."

Rowland's words proved prophetic, although he probably was as surprised as anyone by the source of the new calls for cuts. On March 2, 1989, just 2 months after the Montreal Protocol had gone into effect, environmental officials from the 12-nation European Community agreed to much more stringent restrictions. The group's environment ministers agreed to completely eliminate CFC production and consumption by the year 2000 and to reach an 85% cut as soon as possible. It was ironic that the very governments that had slowed down progress toward CFC cuts earlier in the decade and that had been accused by the United States of selfishness in this respect were the ones that were now leading the cause. "It's a global problem," said French environment minister Brice Lalonde at the meeting. "It's important to get the rest of the world behind us."

The next day, President Bush responded with a declaration of his own, calling for a ban of CFCs by the end of the century, provided safe substitutes could be found. His qualifier about substitutes was a big one, however. As Thomas Burke of the environmental group Green Alliance remarked of the situation, "There's always a big gap between the rhetoric of government and the performance of government."

On March 5, the Soviet Union, China, and India muffled world optimism somewhat by declaring that they were against additional CFC cutbacks. Substitutes for CFCs would prove much more expensive, they said, and they and other countries—particularly in the Third World—simply could not afford it. The Soviet Union went so far as to question the necessity of further CFC legislation, an incredible challenge after two decades of growing ozone awareness.

By the end of the 1980s, the consequences of further delay in eliminating CFCs were too serious to overlook. Holes in the ozone layer were already developing at the South and North poles. In a worst-case scenario, these holes could widen and spread to lower latitudes as a result of positive-feedback processes discussed earlier. Despite these frightening developments, governments continued to haggle over politics and economics. The Montreal Protocol was too weak; even with its provisions atmospheric chlorine concentrations would continue to soar throughout the 21st century. As the 1980s came to a close, it was time for governments and citizens alike to recognize the need for immediate action to save the ozone layer.

The ozone crisis was publicized on Earth Day, April 22, 1990, by a hot air balloon in Madrid, Spain, that proclaimed, "Let's Save the Ozone Layer" and "Stop: C.F.C." The balloon was sponsored by members of the environmental group Greenpeace.

chapter 7

A C T I N G   F O R   T O M O R R O W

The ozone crisis continues. Spray cans have been virtually eliminated, national and international legislation has been enacted to diminish the emission of CFCs into the atmosphere, but they are still being emitted, and many CFCs emitted in the past have yet to exert their effect on the upper regions of the atmosphere. Worldwide phaseouts of CFCs are of course the final answer, but in the meantime it is important for people to recognize how CFCs are being released today and how these emissions can be reduced.

The main use of CFCs in the world today is as a coolant in refrigerators and air conditioners. It is not crucial for people to have spray cans with which to deodorize themselves, but it is important to keep food cool and fresh. Air-conditioning, although essentially a luxury, is something many people are unlikely to willingly do without. Substitute refrigerants are being developed, but until they can be implemented the impact of systems now in use must be softened.

One might think that refrigerators and air conditioners would be fairly safe uses for CFCs. After all, there is no real reason why the CFCs need to ever leave the system, and if they remain inside the machine, they cannot hurt the ozone layer. Unfortunately, most of these products lose about 30% of their CFCs due to routine leakage. The rest eventually escapes to the atmosphere after the unit is junked. Immediate regulation of disposal and leakage restrictions would help to alleviate this problem. A simpler and potentially potent measure would be for individuals to make an effort to use as little air-conditioning as possible until good CFC substitutes are available. This policy would have the biggest impact if applied to automobile air conditioners, because they leak more and are scrapped more often than building units. Automobile units release hundreds of thousands of tons of CFCs into the air each year and, according to some experts, constitute almost 20% of total CFC emissions.

Another immediate measure is to recycle CFCs. Imperial Chemical Industries PLC (known as ICI), the leading British chemical company, has already announced that it would like to do just that, using the same equipment with which the company cleans CFCs at the end of the production process. This would essentially involve removing contaminating oils and water. However, it is still uncertain whether recycling would be more or less expensive than simply replacing the used CFCs. Another problem is that the European Community classifies used CFCs as Special Waste, which means that expensive and time-consuming precautions must be taken with them. ICI says the classification is ridiculous for nontoxic wastes like CFCs. Certainly any unnecessary red tape should be eliminated in order to encourage CFC recycling.

There are two other major uses for CFCs today: as foam-blowing agents and as industrial cleaning solvents. Both of these applications are, at least to some extent, unnecessary.

Among the best-known examples of CFC-blown foam are the plastic containers used in fast-food restaurants. Although a convenient insulating packaging for hot foods, these containers are not absolutely necessary and are clearly harmful. Although CFCs do not leak much from the foam once it is manufactured, the foam-blowing process itself releases large quantities of CFCs. In response to public pressure, the McDonald's Corporation recently announced that it would discontinue use of the plastic cartons. Whether this was done for good public relations or out of concern for the environment, the end result will be less ozone destroyed.

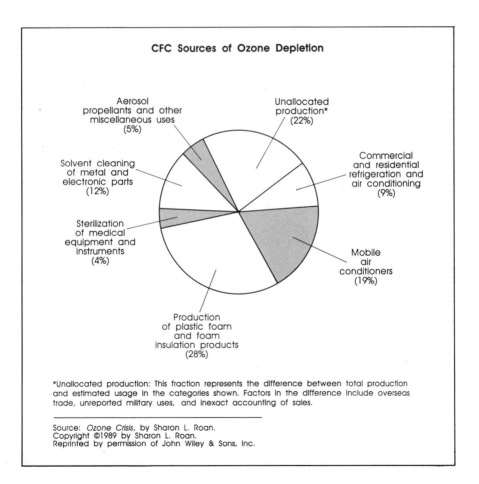

CFC Sources of Ozone Depletion

Aerosol propellants and other miscellaneous uses (5%)

Unallocated production* (22%)

Solvent cleaning of metal and electronic parts (12%)

Commercial and residential refrigeration and air conditioning (9%)

Sterilization of medical equipment and instruments (4%)

Mobile air conditioners (19%)

Production of plastic foam and foam insulation products (28%)

*Unallocated production: This fraction represents the difference between total production and estimated usage in the categories shown. Factors in the difference include overseas trade, unreported military uses, and inexact accounting of sales.

Source: *Ozone Crisis*, by Sharon L. Roan.
Copyright ©1989 by Sharon L. Roan.
Reprinted by permission of John Wiley & Sons, Inc.

One solution to the problem posed by chlorofluorocarbons is recycling. CFCs from refrigerators and air conditioners can be recovered using machines such as this portable device made by Refrigerant Recovery Systems of Tampa, Florida.

CFCs are also not indispensable as industrial solvents. In 1987, the Digital Equipment Corporation in Massachusetts announced that it had already begun using water-based solvents in place of CFC-based systems.

Individuals can exert pressure on other companies to follow examples such as these. Although businesses are slow to act for the environment when large costs are involved, they will respond to pressure from consumers who threaten to take their business elsewhere.

Virtually everyone now seems to accept the fact that CFCs need to be eliminated eventually. The important question in the long run is what to use in their place.

The major chemical substitutes for CFCs in the near future seem to be *HFCs*, or hydrofluorocarbons, and *HCFCs*, or hydrochlorofluorocarbons, which are considered "ozone-friendly" fluorocarbons. These compounds are much more reactive than CFCs and therefore usually break down in the troposphere before they can rise to the stratosphere and damage the ozone layer. Their higher reactivity results from the fact that each molecule in an HFC or HCFC compound has at least one hydrogen atom that can easily be replaced by reactive molecules in the troposphere. CFCs' problematic inertness stems from their complete *halogenation*. This means that each of their carbon atoms is strongly bound to chlorine or fluorine (halogens); HFCs and HCFCs, conversely, have weaker bonds to hydrogen.

Because of their low-altitude decomposition, HFCs and HCFCs were quickly named by industry as the rightful heirs to the refrigeration throne. Du Pont had spent $240 million on their development by the end of 1990 and expects to invest a total of $1 billion by the end of the century. The British company ICI has spent more than £100 million ($200 million) on 2 plants that will produce 100,000 tons of the substitutes annually by 1992 (worldwide industry currently produces about 1 million tons of CFCs each year). Industry officials expect HFCs and HCFCs to constitute 40% of the current CFC market by early in the next century. The rest of the demand will be filled by nonfluorocarbon substitutes still being designed and by recycling.

There are, however, problems with HFCs and HCFCs. As Eric Banks, professor of fluorine chemistry at the University of Manchester Institute of Science and Technology, put it, "Industry is rushing headlong into production of these new chemicals, in almost complete ignorance of their degradation by-products or their effects in the biosphere." In other words, these chemicals may not do much harm to ozone, but they may have other adverse effects, unimaginable now, just as the CFC-ozone problem was unimaginable in the 1930s and 1940s.

So even as industry pours money into new chemical plants, it and government agencies are funding research to see if these new substitutes are really safe. In 1990, the Alternative Fluorocarbon Environmental Acceptability Study, which includes 15 of the leading CFC producers, decided to spend $6 million to study the environmental effects of HFCs and HCFCs. The European Community has allocated £4.3 million (more than $8 million) for similar research. One effect that will be monitored closely is just how much these new substances might contribute

An engineer tests substitutes for CFC solvents at a Du Pont facility in Wilmington, Delaware. New alternatives to chlorofluorocarbons are less damaging to the ozone layer but may present other, as yet unknown, hazards to the biosphere.

to the greenhouse effect; it is already known that they are indeed greenhouse gases.

Aside from possible environmental dangers, HFCs and HCFCs are much more expensive to produce than CFCs. Du Pont estimates that it will cost almost $400 billion to replace CFCs with their less halogenated cousins. Almost all of this figure would be for refrigeration—particularly in developing countries, where demand for refrigeration is increasing dramatically.

Fortunately, refrigeration technology featuring nonfluorocarbon coolants may soon be available. As Debora MacKenzie wrote in the June 1990 issue of *New Scientist*, "It is possible to abandon not only CFCs but their relatives as well, and to do so cheaply. The only financial assistance needed will be investment credit to make the change; the substitutes will then pay for themselves." She was referring to the use of coolants such as propane and even water.

Refrigerators using propane gas or water are already operating successfully in laboratories in Europe, and tests indicate that they may be even cheaper and more energy efficient than standard refrigerators. Nevertheless, these new systems have been completely ignored at international conferences dealing with the phaseout of CFCs, according to proponents of propane and water systems because the industries being consulted are experts only in CFC production, not refrigeration.

The new propane technology is also being held back because of an old assumption in the refrigeration industry that propane is not a wise choice for a coolant. Early refrigerators used as much as 11 liters of coolant, rendering flammable propane a dangerous option. When nonflammable CFCs were invented, the use of propane was abandoned. Modern refrigerators, however, use only one-hundredth as much coolant, or about 10 milliliters, making the flammability issue immaterial. In fact, refrigerators designed specifically for propane would use only 5 milliliters (the same amount as a cigarette lighter), not to mention about 10% less energy than a standard CFC model.

At the moment, legal restrictions stand in the way of developing propane refrigeration. In the United States, several gallons of propane can be stored indoors in a portable container, but in a permanent container such as a refrigerator even the small amount required is subject to extraordinary safety regulations. In Britain, antiquated laws prohibit all flammable refrigerants. Pressure from citizens and their representatives in government could remove these stumbling blocks.

MONTREAL REVISED

It looks as though pressure of some kind will always be necessary before industry will develop environment-friendly technologies. This was the idea in mind when the Montreal Protocol nations met in London in late June 1990 to try to strengthen the measures of the original treaty. They knew that ozone depletion would continue for centuries without a stricter timetable enforced on the CFC producers. The new agreement, reached late on the night of June 29, did just that.

The most significant result was that all parties agreed to move from a 50% reduction of CFCs to a 100% elimination by the end of the century. They also set deadlines of 1995 for a 50% cut and of 1997 for an 85% cut. Also, other ozone-damaging halogen compounds were now included in the phaseout. Less harmful HCFCs and HFCs were not officially included, but a resolution was passed that should eventually provide for the elimination of these substitutes between the years 2020 and 2040.

A crucial development at this meeting was that China and India, major new producers of CFCs and previously unwilling to sign the protocol, agreed to the phaseout. The key to their change of heart was that the signing countries agreed to set up a special fund, worth $240 million over the first 3 years, to help these and other developing countries adapt to new substitute technologies. This had been a big stumbling block until just weeks before the meeting, when the United States finally agreed to help finance the fund.

This component of the new agreement helps solve one of the biggest problems associated with phasing out CFCs. Third World countries had been without the benefits of cheap refrigeration while the industrialized nations were taking

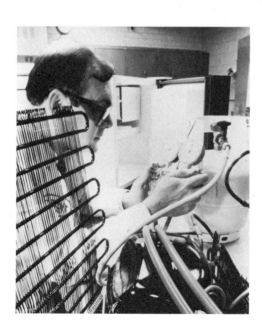

A Du Pont engineer inspects a refrigerator containing one of the company's new non-CFC coolants.

advantage of CFCs. Then, just as Third World nations were about to develop the industry themselves, the rest of the world decided that it must be scrapped. Somewhat understandably, these developing nations were not prepared to give up their rights to cheap refrigeration without financial help. Now the wealthier countries will provide help for the transition to alternative systems, and before long new technologies should be available that will be affordable for all.

INTO THE FUTURE

Twenty years after the discovery that CFCs were destroying the protective ozone layer, the international community is finally acting to eradicate them. By the end of the century, they will probably no longer be an issue. The consequences of their use, however, remain.

In April 1991, the Environmental Protection Agency made the startling announcement that ozone depletion is occurring faster than had been anticipated throughout the Northern Hemisphere temperate zone—that is, over North America, Europe, the Soviet Union, and much of Asia. Ozone loss in this region averaged 4.5% to 5% over the last decade, according to research by scientists at the Goddard Space Flight Center. Depletions have worsened elsewhere around the globe as well. The EPA now calculates that the resulting increase in penetrating ultraviolet radiation will cause 12 million skin cancer cases and more than 200,000 deaths from cancer over the next 50 years in the United States *alone.*

Will ozone depletion percentages continue to climb? And what will happen in coming years to the ozone holes over the

Antarctic and Arctic? How will they affect the complex systems that determine the earth's global climate? How will a continued global warming trend affect the damaged ozone layer, and vice versa? And how much of an effect is yet to come from the millions of tons of CFCs already released that are still making their way up toward the stratosphere?

No matter what the answer to these questions, humankind should already have learned an important lesson from the ozone problem. There is nothing permanent about the earth, much less about any of the individual systems that make up its biosphere. Moreover, human activity, long considered in-significant next to the enormity of these systems, is now known to have a measurable and harmful effect on them.

As a result of industrial progress, an entirely different world could emerge in the next hundred years, one with a colder stratosphere, warmer troposphere, and much more dangerous ultraviolet radiation reaching the ground. Disastrous climate changes and millions of new skin cancer cases could occur. These changes, if they happen, will affect the earth in the 21st and 22nd centuries, when today's children become grandparents and when their grandchildren become grandparents. The only way to prevent or at least mitigate these consequences is to act now.

THINGS EVERYONE CAN DO NOW

∘ Do not buy household products, including certain hair sprays, spot removers, and solvents, that contain ozone-destroying chemicals such as CFCs and methyl chloroform.

- Purchase non–halon containing fire extinguishers.
- Avoid using polystyrene foam products—food and drink containers, insulation materials, and packaging—that contain CFCs. Pressure companies that use these products (such as fast-food and mail-order businesses) to switch to safer alternatives. Methods of persuasion include writing letters to company officials, refusing to buy unnecessary packaging, and boycotting when other strategies fail. Many companies have already responded to such consumer pressure.
- Ask dry cleaners to stop using CFC solvents or—to avoid dry cleaning altogether—buy clothing made from washable fabrics, such as cotton.
- If your home or automobile air conditioner is serviced, ask that the CFC-containing coolant be drained into a closed container rather than evaporated; also request that it be recycled. Replace car air conditioner hoses every three years to prevent CFC leakage.

APPENDIX: FOR MORE INFORMATION

Environmental Organizations

Chemical Manufacturers
  Association
2501 M Street NW
Washington, DC 20037
(202) 887-1100

Environmental Action Coalition
625 Broadway
New York, NY 10012
(212) 677-1601

Environmental Action Foundation
1525 New Hampshire Ave. NW
Washington, DC 20036
(202) 745-4879

Environmental Defense Fund
257 Park Avenue South
New York, NY 10010
(212) 505-2100

Friends of the Earth
530 7th Street SE
Washington, DC 20003
(202) 544-2600

Greenhouse Crisis Foundation
1130 17th Street NW/630
Washington, DC 20036
(202) 466-2823

Greenpeace USA
1436 U Street NW
Washington, DC 20009
(202) 462-1177

Natural Resources Defense
  Council
40 West 20th Street
New York, NY 10011
(212) 727-2700

World Resources Institute
1709 New York Avenue NW
Washington, DC 20036
(202) 638-6300

Worldwatch Institute
1776 Massachusetts Avenue NW
Washington, DC 20036
(202) 452-1999

Government Agencies

Environmental Protection
  Agency (EPA)
401 M Street SW
Washington, DC 20460
(202) 382-2090

United Nations Environment
  Program
North American Liaison Office
Room DC 2-0803
United Nations, NY 10017
(212) 963-8093

FURTHER READING

Dotto, Lydia, and Harold Schiff. *The Ozone War*. New York: Doubleday, 1978. The definitive account of the 1970s ozone controversy.

Environmental Information Exchange Citizen Guide. *Protecting the Ozone Layer: What You Can Do: A Citizen's Guide to Reducing the Use of Ozone Depleting Chemicals*. New York: Environmental Defense Fund, 1988.

Firor, John. *The Changing Atmosphere*. New Haven: Yale University Press, 1990.

Fisher, David E. *Fire and Ice*. New York: HarperCollins, 1990. An easy-to-understand discussion of ozone depletion, global warming, and other impending global disasters.

Graedel, Thomas E., and Paul J. Crutzen. "The Changing Atmosphere." *Scientific American* (September 1989): 58–68.

Gribbin, John. *The Hole in the Sky*. New York: Bantam Books, 1988. A comprehensive treatment of the scientific and political aspects of the ozone issue.

MacKenzie, Debora. "Cheaper Alternatives for CFCs." *New Scientist* (June 30, 1990): 39–40.

Miller, Alan S., and Irving M. Mintzer. *The Sky Is the Limit: Strategies for Protecting the Ozone Layer*. Research Report No. 3. Washington, DC: World Resources Institute, November 1986.

Molina, Mario J. "The Antarctic Ozone Hole." *Oceanus* (Summer 1988): 47–52.

Oppenheimer, Michael, and Robert H. Boyle. *Dead Heat: The Race Against the Greenhouse Effect.* New York: Basic Books, 1990.

Roan, Sharon L. *Ozone Crisis.* New York: Wiley, 1989. A month-by-month narrative of the political ozone wars, from 1973 to 1989.

Sayed, Sayed Z. el-. "Fragile Life Under the Ozone Hole." *Natural History* (October 1988).

Schneider, Stephen H. *Global Warming.* San Francisco: Sierra Club Books, 1989.

Shea, Cynthia Pollock. *Protecting Life on Earth: Steps to Save the Ozone Layer.* Worldwatch Paper 87. Washington, DC: Worldwatch Institute, December 1988.

Tickell, Oliver. "Up in the Air." *New Scientist* (October 20, 1990): 41–43.

UNEP. *The Ozone Layer.* UNEP/GEMS Environment Library No. 2. Nairobi, Kenya: UNEP, 1987. Available from UNEP, 1889 F Street NW, Washington, DC 20036.

Young, Louise B. *Sowing the Wind: Reflections on the Earth's Atmosphere.* New York: Prentice-Hall, 1990.

Readers can gain a better understanding of both scientific and political aspects of the ozone issue by searching out past, present, and future issues of the magazine *New Scientist.*

GLOSSARY

AAOE   The Airborne Antarctic Ozone Experiment (August–October 1987).

AASE   The Airborne Arctic Stratospheric Expedition (spring 1989).

aerosol   Any fine mist of droplets; not necessarily artificial or related to spray cans.

albedo   The reflectivity of the earth, or how much sunlight is reflected back into space rather than absorbed.

Antarctica   The continent on which the South Pole lies; more than 5 million square miles in area and almost entirely covered by ice.

arctic   Of or relating to the geographic area north of the Arctic Circle (latitude 66°33′ north) and including the North Pole.

catalyst   A substance that increases the rate of a chemical reaction without being consumed in the process.

chlorofluorocarbons (CFCs)   Ozone-destroying chemical compounds developed in the 1930s as refrigerants and later used in spray cans and other products.

computer models   Computer-simulated representations of an object of study, such as the complex systems that determine climate; used to study the potential impact of various changes, such as ozone depletion.

coolant   Liquid used in refrigerators and air conditioners to absorb heat.

dissociation   The separation of a molecule or chemical compound into smaller parts. See also **photodissociation**.

Freon   The first marketed CFCs—trichloromonofluoromethane and dichlorofluoromethane; used mostly as coolants.

Gaia   A concept, developed by scientist James Lovelock, that envisions the earth as an organism with self-regulating processes that maintain a stable global environment able to support life.

greenhouse effect   The trapping of infrared radiation in the earth's atmosphere by gases such as carbon dioxide and methane, resulting in higher temperatures.

hydrochlorofluorocarbons (HCFCs) and hydrofluorocarbons (HFCs)   Substitutes for CFCs that break apart more easily in the troposphere and thus do not usually rise intact to the stratosphere and destroy ozone.

nanometer (nm)   One billionth of a meter.

nitrous oxide   ozone-destroying gaseous compound that also contributes to the **greenhouse effect**.

NOZE   The National Ozone Expedition, which took place in Antarctica (August–October 1986).

photodissociation   The breakdown, caused by ultraviolet radiation, of substances such as oxygen molecules or CFCs.

polar stratospheric clouds (PSCs)   Enormous thin clouds of ice crystals that form in the Arctic and Antarctic in the winter months and that trigger dissociation of CFCs. *See also* **surface catalysis**.

SBUV   The Solar Backscatter Ultraviolet measurer; used aboard the *Nimbus 7* satellite to record ozone concentrations.

spectrophotometer   Instrument used to measure ozone in the atmosphere by detecting solar radiation.

stratosphere   The ozone-containing layer of atmosphere above the troposphere, between approximately 7.5 and 30 miles altitude.

surface catalysis Process that takes place on the surface of tiny particles of ice in **PSCs** whereby CFC molecules dissociate more quickly.

TOMS Total Ozone Mapping Spectrometer; used aboard the *Nimbus 7* satellite.

troposphere The lowest region of the atmosphere, in which most weather takes place; extends from the surface of the earth up to an average altitude of 7.5 miles.

ultraviolet light Radiation with wavelengths measuring between 280 and 400 nm; UV-B radiation (280 to 320 nm), normally absorbed by the ozone layer, is harmful to life on earth.

INDEX

PICTURE CREDITS

AP/Wide World Photos: pp. 21, 24, 58, 74, 90; The Bettmann Archive: pp. 45, 62; Du Pont Corporation: p. 93; Dr. David Hofmann—NOAA: p. 56; Courtesy of Mario J. Molina: p. 41; NASA: pp. 12, 15, 48, 53, 79; Refrigerant Recovery Systems, Inc., Tampa, FL: p. 88; Reuters/Bettmann: pp. 36, 84; Runk/Shoen-Erger from Grant Heilman: p. 42; Illustrations by Gary Tong: pp. 17, 19, 33, 50, 81, 87; University of California, Irvine, California: p. 31; UN Photo 151–869/John Isaac: p. 69; USDA: p. 71; USDA Photo by June Davidek: p. 67

A B O U T   T H E   A U T H O R

MARSHALL FISHER is a freelance writer living in Boston. He was most recently a writer and researcher at *Earthwatch* magazine and before that worked as a tennis pro in Munich and as a sportswriter in Miami. He received a B.A. in English from Brandeis University and an M.A. in English and creative writing from the City College of New York. His work has appeared in *Kether*, the *South Dade News Leader*, *Earthwatch*, and *Sources*.

A B O U T   T H E   E D I T O R

RUSSELL E. TRAIN, currently chairman of the board of directors of the World Wildlife Fund and The Conservation Foundation, has had a long and distinguished career of government service under three presidents. In 1957 President Eisenhower appointed him a judge of the United States Tax Court. He served Lyndon Johnson on the National Water Commission. Under Richard Nixon he became under secretary of the Interior and, in 1970, first chairman of the Council on Environmental Quality. From 1973 to 1977 he served as administrator of the Environmental Protection Agency. Train is also a trustee or director of the African Wildlife Foundation; the Alliance to Save Energy; the American Conservation Association; Citizens for Ocean Law; Clean Sites, Inc.; the Elizabeth Haub Foundation; the King Mahendra Trust for Nature Conservation (Nepal); Resources for the Future; the Rockefeller Brothers Fund; the Scientists' Institute for Public Information; the World Resources Institute; and Union Carbide and Applied Energy Services, Inc. Train is a graduate of Princeton and Columbia Universities, a veteran of World War II, and currently resides in the District of Columbia.